the naked truth

the naked truth

YOUNG, BEAUTIFUL, AND (HIV) POSITIVE

marvelyn brown

with courtney e. martin

AMISTAD
An Imprint of HarperCollins*Publishers*

This is a work of nonfiction: all the events depicted are true and the characters are real. The events, conversations, and experiences detailed herein have been faithfully rendered as I have remembered them, to the best of my ability. Some names, identities, personal characteristics, and circumstances have been changed in order to protect the privacy and/or anonymity of the various individuals involved.

HarperCollins books may be purchased for educational, business, or sales promotional use. For information, please e-mail the Special Markets Department at SPsales@harpercollins.com.

FIRST EDITION

Designed by Susan Yang

Library of Congress Cataloging-in-Publication Data has been applied for.

ISBN 978-0-06-156239-6

17 18   OV/RRD   20 19 18 17 16 15 14 13 12

For my nieces Diamond and Jamiya
and my nephew Jamarius.

For anyone who has ever been
isolated, disowned, or felt they had to lie
to be loved, this is for you.

prologue

"What is wrong with this child?"

I heard my momma's voice even before I opened my eyes. I heard the voices of my aunt, my uncle, my grandmother, even my daddy, who I hadn't seen in the last year. My situation had to be serious for him to come out.

A flood of light almost blinded me when I finally managed to lift my heavy eyelids. Everything was the bright white and metallic colors of a hospital room. The air smelled antiseptic, like Lysol and plastic. It was the smell of death.

I looked down at my body, shrunken from my previous athletic build—five feet four inches, 128 pounds, pure muscle—to a skeletal 109 pounds, all skin and bones. There were little white circles stuck all over my body, scary wires coming from each one like roots growing out of a potato. My mouth felt dry, my throat sore. I could feel the naps forming in my hair.

I looked up at the ceiling while my brain reached back into my memory, searching for the last thing I could remember. Being pushed fast through the intensive care unit in a wheelchair, dry gagging because I was dehydrated, sharp pains slashing through my stomach, my eyes filled with water. I could not hold my head up. That was it. And before that, walking down my auntie's stairs and then passing out. Fainting. Unconscious.

I closed my eyes again, hoping my family wouldn't notice that I had woken up.

When I opened them again, I thought it had to be nighttime. I was surrounded by a curtain, and everyone had gone home except for my daddy, who slept slumped over in a chair in the corner. I studied his face, slack with sleep. He looked old, older than his forty-four years. He had bags under his eyes—a signature family trait—and a mustache. His hair was short; at this point the gray was dominating the black. No matter the situation, it was always a pleasant feeling to be reassured by my dad's love for me and to know that he cared.

"Mom must hate that he's here," I thought to myself. My parents had been divorced since I was three years old, and my mom didn't think much of my dad—who'd basically vanished after that. Then I heard her voice echo in my head: "What is wrong with my child?" I didn't have a clue.

I had started feeling tired a couple of weeks earlier, but I'd figured it was because I was working so much—one shift at my aunt's day care and then one at a pizza parlor every day; both paid only seven dollars an hour. I was trying to save up money for college tuition. I had to get up at six in the morning, and my day usually didn't end until past midnight. As tired as I was, I usually found time to visit my man after work.

But as the days wore on, I kept feeling worse. I didn't even have the energy to do my hair and take a shower. It wasn't like me to roll down the stairs in the morning without putting myself together or at least throwing on a cute hat. Since growing out of my tomboy phase in high school, I'd always put care into my appearance, but I just felt too damn tired to even make the effort. My appearance worried my aunt, and after a day of me sleeping for fifteen hours straight and turning down my favorite Hamburger Helper meal, she called my mom.

My mom thought my aunt was crazy—"Sick? That girl's fine. You can't tell all that from her appearance!"—but she took me to Vanderbilt University Hospital just to prove herself right. She thought I might be pregnant.

As soon as we got to the hospital, she told the doctor, "Give her a pregnancy test," as she glared at me judgmentally.

"I ain't pregnant," I said, without even looking in her direction. I was on my cell with my friend Cortney figuring out where we were going to go out that weekend.

My momma and I almost never got along, so even though I sort of suspected I was pregnant myself, I shook my head like she was out of her mind. "Stay out of my business," I said under my breath. Ever since I was a little girl, my momma had been warning me, "Just don't get pregnant." It was as if she thought that was all the sex education I needed. Why was she trying to have the sex talk now? It only embarrassed me. I remember thinking, "We never talk about this. Why are we talking about it *now*?"

When the nurse said the pregnancy test was negative and that I had a twenty-four-hour virus, I was relieved, though still a little bit worried about why I was so tired. "Told you I wasn't pregnant. Are you happy now?" I asked my mom with attitude.

I crashed at my aunt's house after that. I usually slept there, at my sister's house, or at my grandmother's. The next day I fainted on the way down the stairs, the final straw that got me hemmed up like this in the hospital.

I didn't know how many days had gone by or what the doctors were learning from all these wires sticking out of me, but frankly, I didn't care. You see, I was only nineteen, but life had already been hard enough for me—and I was well along the path of self-destruction.

As I lay there in that hospital bed, helpless and wired up, I didn't have a feeling in the world. It was as if I was saying to myself, softly, like I used to talk to the toddlers at the day care during nap time: "Don't fight it. Just close your eyes and go to sleep. It will be much easier."

I had considered suicide before but always stopped myself because I knew that taking your own life could send you straight to hell. But lying there with the heat of hell already in my head—a 106-degree temperature with no sign of cooling—I realized that this was my chance to pass away quietly without enduring the wrath of God. It wouldn't be my fault. I would just go to sleep . . .

As fate would have it, that was not the moment of my death but the beginning of the rest of my life.

You see, I was born Marvelyn Brown, a little black girl in Nashville, Tennessee, on May, 7, 1984, but I wouldn't understand the significance of my existence until it was threatened a couple of decades later. At just nineteen years of age, I was unsure what was killing me and, even more disturbing, unwilling to fight.

This is the story of my still young, wild, and, at times, reckless ride of a life. It is a cautionary story of how I discovered self-love, self-respect, and responsibility and began traveling the globe speaking a truth in hopes of saving the lives of others. It is the story of my marvelous life.

My childhood modeling headshot.

chapter one

As a child, I adored my mother, but I was definitely a daddy's girl. I used to love to just sit and watch my father play his favorite game of all time: Pac-Man. As the little yellow man whisked across the screen, my daddy would go from happy to overexcited to sad to angry—all in a matter of seconds. I would patiently wait for Inky, Pinky, Blinky, or Clyde to eat him up. Then he would yell and curse at the screen, and I would laugh until my stomach hurt.

Unlike so many other black fathers, who disappeared as fast as Pac-Man, my dad, Marvin, was the present parent in those early days. He would take my sister, Mone't, and me to kids' movies, even the ones he hated watching, and nap until the credits rolled, at which point he would clap enthusiastically as if he had been awake for the whole thing.

I loved going to the grocery store with my dad because he would always buy me yummy treats, even if it was right before dinner. We would sneak and eat candy, disposing of all the wrappers before we got home so my mom, Marilyn, wouldn't find out. My father and I would walk into the house with a secret, shared smirk on our faces. Those smirks disappeared when I went to the dentist and our cover was blown. I had eight cavities. That night my father slept on the couch and I was on grocery-store suspension.

Back then I didn't sense the tension between my parents, though I could tell that my mom—an engineer and a labor organizer—was far more on top of things than my dad, who fixed copy machines for Xerox. Neither of my parents came from money. This was the South, after all. My parents' was really the first generation of southern blacks that had a chance at a decent life. My mom wanted us to be a "successful" family, which in her mind meant all work and no play.

My mom was the two S's, *strict* and *serious*. She was so driven that I don't remember her ever letting her hair down. She was always nagging my dad about this or that, scolding Mone't, or me (more often me) for not acting the way she wanted us to, or ordering us around. It seemed like we could never do right, never be enough for her.

From the time I could walk and talk, I was involved in a million activities—dance, modeling, swimming, track. If my mom saw a sign-up list, you'd better believe my name was first on it. She didn't care what it was. She just wanted me busy, busy, busy. It wasn't just that she thought it would keep us out of trouble. She also thought it would help us achieve later on in life. And that's pretty much all my mom cared about: success.

My dad and I were sitting on the couch hanging out one afternoon when my mom came home from work and brought a dark cloud of unhappiness with her. I was only five, but I was already a master at picking up on my mom's emotions. I can still remember the visceral feeling I had when she was fed up.

"Marvelyn, your dad and I need to talk. Get in the other room."

I scurried away, hearing that she meant business, and she closed the door behind me. I remember just looking at that closed door, wishing I could understand how to make the problems between my mom and dad all better. I got down on my knees and put my ear to the door, hoping to hear, then wishing I hadn't. My mom was yelling at my dad, calling him "no good," ordering him to leave. I began to shake, then I broke down crying. I couldn't believe this was happening. Why weren't any of us ever good enough for my mom? Why did my dad's fun-loving nature make her so mad? Why did she feel so worried about us "making it" all the time?

I know now that my mom was dealing with a whole lot of drama I didn't have the first clue about. For starters, my dad was addicted to drugs. I didn't know this at the time. Hell, I didn't know it until a couple of years ago when I finally asked. Those kinds of things often get covered up in families—especially in southern families. We don't like to air our dirty laundry.

As it turned out, there was a lot of it. Not only was my dad addicted to gambling, drugs, and alcohol, but he had other kids. Years later I would ask my mom, "Why did you marry Dad if he had other kids?"

She shot back, sounding characteristically stern, "You think I would have married him if I'd known about those kids? Please, Marvelyn, I didn't have a clue." Of course, I realized—that would have ruined her image of the perfect "successful" family.

No matter how much my mom tried to keep it together, the image was shattered anyway. My parents got a divorce, and my mom gained full custody of my younger sister and me, along with the house and cars. Our family went from the ideal, middle-class picture to a single-parent household where the mom is trying to hold it down and the dad just makes cameo appearances. From then on I heard from my dad a few times a year at most, usually around the holidays and my birthday. One day when arriving home from school I realized that the answering machine was blinking with a message. I ran over to press *play* and heard my dad wishing me a happy birthday. I was so excited to hear from him and get the birthday wishes that I hardly noted that he was five days late.

Once my dad left, my mom's focus on Mone't and me got even sharper. Even though she was working all the time, she managed to be highly involved at our schools. She was one of the only non-white mothers on the PTA, and she made sure to handpick our teachers every year. My classmates thought my mom was so cool because a lot of their mothers were not as active, but I felt like it was an invasion of my privacy and just added to the pressure I felt.

We were her works-in-progress. I hated how much she pushed me, and even more, I hated that she had sent my daddy away. I didn't understand why she made him leave and was sure that his continued absence was mostly her fault. Through a child's eyes, I saw only a loss of fun, play, love. The drugs were invisible.

Money troubles were concealed from me too. In fact, my mom was struggling. With the loss of my dad's income, she was forced to work outrageous twelve-hour night shifts to keep food on the table and a roof over our heads. It seemed like she was either at work, asleep, or ordering us around; she often joked that she'd just had kids so they could do the work around the house. Sometimes I wasn't sure she was joking.

I resented that she ordered us around, not just because I didn't like the work but because she used guilt to manipulate us. She would reference the Bible—"'Honor thy mother,' Marvelyn." She knew I had a close relationship with God. I decided that I was doing the Christian thing by absorbing everything she said even though every night I would cry myself to sleep.

My mom and my sister even used to gang up on me and call me names like Cinderella and Dumbelina. I often compared myself to children that I would see on the late-night television specials in Third World countries who had plenty of love and support but no money for food. I felt like I was the total opposite—I had all the food I needed but was insatiably hungry for love.

I tried to do my part to help out. I learned how to cook from watching television and would make Mone't and myself heaps of steaming scrambled eggs. When we got sick of eating eggs every day, I graduated to spaghetti and eventually to fried chicken.

Babysitters were constantly cycling in and out of our house. We must have had fifty over the course of ten years or so. Too exhausted to care, my mom was not particular about who watched my sister and me. She had only one rule—the babysitter had to be at least twelve years old. To this day, when I visit back home I run into people who ask if I remember them. When I give them a blank stare they exclaim, "I babysat you when you were little!" Yeah, you and the rest of the 'hood, I want to answer. But I just smile instead.

Having all those babysitters might sound a bit frightening, but it was actually a good thing for me. I was exposed to opinions different from those of my domineering, exhausted mom. They were all young and enjoying life. They didn't yell at us for playing or see it as a waste of time. When they were around I didn't have to be responsible; I could just be a bona fide kid.

My favorite babysitter was my older half-sister on my dad's side. Tab would watch us with some of her high school friends, and boy, did we have a ball. A few of our many quests were stealing my mom's car to go to the store and buy Now and Laters, and borrowing eggs from the neighbors, only to break them all running away from a rabid dog. We were always up to something.

Another time Mone't and I got it into our heads that we were going to earn a little money ourselves—young entrepreneurs making Mommy proud. As my older sister slept, Mone't and I cleared out the garage to make room for our new project, a playroom. We put a lawn mower, tools, old chairs, and several knick-knacks into the front yard. We made FOR SALE signs with our crayons and watercolors and put them up around the block. As we were sitting on the grass in lawn chairs, drinking lemonade, my older sister came racing from the house screaming like a madwoman, with a belt in her hand. We dropped that lemonade and ran for it, scared for our lives (and our butts), but by the time she caught up with us, she could do nothing but laugh. We were scrappy, even then.

By the time I got out of elementary school it was becoming pretty clear that Mone't and I had very different talents. I was the athletic one and she was the smart one. I didn't mind this identity. After all, I was the one who brought home the blue ribbons from the field day and ran around the neighborhood with the boys. I considered myself a tomboy and liked it that way. There were too many females in my house as it was. I missed my dad, and being around the neighborhood boys reminded me of his goofy sense of humor.

My mom liked my blue ribbons, but she also wanted me to be a "pretty" girl. She had me modeling, dancing, acting, and singing. I did a couple of commercials, worked as an extra in two movies, and appeared in several print ads. I was always eager to get out of those dresses and back to the streets to find "my boys" in the neighborhood.

Though I hammed it up for the camera, worked hard at the track, and got good grades, it was never enough to make my mom proud. I went to the Junior Olympics every year, and but she never let me forget that without the team, I would have never made it. (I got eliminated early in individual events, because I didn't yet have the confidence to succeed on my own.) When I got B's, she asked why they weren't A's. When I got A's, she asked why they weren't A pluses.

I consistently felt like I was an embarrassment and a disappointment to my mom. I watched longingly as other kids loaded their backpacks up with swimsuits and sunscreen for summer camp. No parents. No schedules. No dresses. With all of my activities, Mom said I didn't have time for that.

When I did have free time, I was hanging with "my boys." Whatever they played, I played; it didn't matter if it was football, baseball, or wrestling. I was never into Barbie dolls or stuffed animals. I didn't do "cute." I loved running and getting hurt and dirty.

I can still hear the sound of my mom's voice, which often met the slam of the door as I would slide in after playing: "Childhood molds the future, Marvelyn Brown! Too much playing around isn't going to lead to anything! You gotta get serious if you want to make something of yourself!"

Whenever she thought we were playing too much, she would invoke the example of the Jacksons: "Michael Jackson worked hard as a child. He didn't have time to play. Now he bought his own amusement park, Neverland Ranch, and he can play in that."

I didn't want to wait until I grew up to play. Unfortunately, that is just what happened.

"We're moving." My mom said it with about as much energy as if she'd said, "Pass the salt." Just entering middle school, I'd been around enough to understand that when she tried to act like something didn't bother her, it usually meant it bothered her a lot.

My mom had been so proud of our home and the neighborhood it was in, so proud of the life she had made for Mone't and me (despite my deadbeat dad). To move was to admit defeat, admit that there were chinks in her armor, admit that she needed others—we were moving in with our grandmother on the other side of town.

It also broke her heart because the move revealed that she was not her own mother's golden child, as she liked to think. She had spent years badmouthing her sister and one of her brothers. They were the screwups; she was the success story. Or at least that's how she liked to paint it. She liked to think of herself as completely independent; now she had to admit that everyone needs someone.

It was a big adjustment for all of us. I switched out of a public school my mother considered not good enough for me to attend a magnet school known for dance, drama, and arts. On paper it seemed like just the right place for me, but in reality it felt wrong in every way.

I was constantly acting out—talking back, getting into fights—because I didn't know how to cope with being the new kid and feeling like I didn't belong. My gang of boys back in our old neighborhood had been replaced by a classroom of unfriendly faces. Other kids made fun of my accent, calling me Oreo because they said I tried to "talk like a white girl." They also made fun of my nappy hair; I wasn't allowed to get it straightened.

My grades began to slip, and before I knew it, I had adopted a blasé, rebellious attitude toward just about everything. It seemed as if the more I acted out, the more the kids liked me. When I smarted off to substitute teachers, I was hero for a day. When I started fights on the playground, I was no longer Oreo but a tough girl.

Soon the principal and the teacher in charge of suspension knew me by name; I thought it was cute, as if I was just playing another role in one of those stupid commercials. When parents of classmates wouldn't let me come to their birthday parties because I'd earned a reputation as a "bad influence," I wondered if things had gone too far. I had just wanted to fit in. Now I was being shut out.

It was around that time that I did something I will never forget. My mom ordered me, as usual, to get her a drink. The anger and frustration just bubbled up uncontrollably inside of me. I felt so powerless when it came to my mom's constant orders and criticism, so sick of not being enough for her.

I filled her cup with Sprite—then added a splash of bleach. It wasn't a conscious choice. It was like that starving part of me was in charge, a part that needed to take back some kind of power. I handed her the cup and walked quickly out of the room. I heard her spit it out and come running after me. I lied and said that I found the cup on top of the kitchen sink, where my grandmother usually had a cup with bleach. But she knew I was lying. I was the talk of the family for a week. My grandmother, who understood the good in me, did not believe that I had done it. Truth is, I didn't want to believe I'd done it either. It made my unhappiness very, very real.

My mom did *not* like my behavior, to put it mildly. She would constantly remind me that I was "going to be nothing in this world" if I didn't change my ways. That was the extent of our "heart-to-hearts." The truth was that if my mom had anything important to tell me, she would communicate it through one of my aunties or my grandmother. At one point, she even sent me to a counselor.

It broke my little heart that my mom couldn't tell me she loved me or have a real conversation with me. It was as if she only spoke in nags and gripes. Looking back now, I see that she did these things—the counselor, asking my aunties to talk to me, even the nagging—because she loved me. But it would have meant so much to me to hear her say that directly, just once.

Mone't and I had never gotten along very well, but my acting out created a bigger abyss between us. My sister, two and half years younger, was still obedient and adjusted to our new lives under Grandmomma's roof with ease. Mone't and my mom had a bond that I envied. I suspected that my mom loved my little sister so much more because her future appeared a lot brighter than mine. Mone't was the well-behaved teacher's pet who made straight A's. She attended a school focused strictly on academics where students were hand-chosen to attend.

Mone't was the brain and I was the brawn. My mom looked at it like this: If my little sister fell and broke her ankle, she still had a chance of succeeding in life; but in my case, the future would be over before it even started.

I realized that she saw potential in my athletic skills, and I ran with it. If that was my only hope of "making it," as my mother seemed to believe, then that's what I would focus on. In seventh grade I began to play more team sports—volleyball, basketball, and soccer—and continued running track. I was naturally gifted at athletics, but mostly I was also fueled by my never-ending desire to earn my mother's love.

When the WNBA was announced in 1996, my mom's dream of me becoming a professional basketball player was born, but I didn't share this dream, and this was another source of conflict between us. I played sports for fun, to keep my mom of off my back and make her proud. Although I was competitive, I did not always play to win. But my only path to success, in my mother's eyes, was athletics (she had given up on me academically).

One day, I discovered something that made me wonder if my mom was also harder on me because she saw herself *in* me. While she and my grandmother were at work, I decided to climb up into the attic and look around. It was full of unlabeled boxes and bags, all covered by a thin layer of dust. I stumbled upon my mom's old high school yearbook and blew the dust off. As I flipped through the pages, I discovered that my mom had been a star athlete herself. I also found faded newspaper clippings that revealed that she had been the leading scorer on her basketball team.

That was the thing about my mom—she was so hard on me, but I knew deep down that it had something to do with her being hard on herself. Being a single mother of two young girls, wanting them to avoid heartache and be successful, caring so much what other people thought—it all added up to a lot of stress for her. And I couldn't blame her. As much as I hated the way she drove me, the way she made me feel like I could never do or be enough, I knew she was just trying to raise me right. She had a strange way of showing love—no hugs, no "I love you," ever—but I knew that she meant well.

With my mother's name being Marilyn and my father's Marvin, you might think that my name, Marvelyn, originated from a combination of their names. My mom once said that she named me Marvelyn because my birth was "a marvelous occasion," but that was because she was being interviewed and wanted to make it sound good. In fact, I was named after one of my mom's high school classmates. Ever since my mom heard the name Marvelyn being called at class roll, she thought the name was beautiful.

I know this is the true story of my name, but still, the sound of my name made me all the more aware that I was one part my mom and one part my dad. The part of me that is fiercely independent and sometimes stubborn, the part that craves success and is unafraid of taking on the world—that is my mom. But my fun-loving side, the Marvelyn at the club and laughing, the Marvelyn who cares deeply about people and isn't afraid to tell them, that's my daddy.

When I look back on my childhood, I am struck by how alone I felt, how resilient I was, how much I suffered because of my mom's outsized, unfulfilled expectations. Neither she nor I had any idea how far I would go, or how the Marilyn part of me would play such a key role in reaching those heights.

But before that, I was going to hit rock bottom.

My nieces Diamond and Jamiya and
my nephew Jamarius when they were little.

chapter two

Putting middle school behind me felt good. High school offered a fresh start, and I planned on making the best of it. I went to a predominantly black school that was known as a "fashion show." I wanted to keep up, especially because it was my first time attending a nearly all-black school. I had to catch up with the language if I didn't want to hear the same old "Oreo" insults I had gotten in middle school.

I jumped into athletics with a vengeance, fully embracing my tomboy identity once again. I had a smart mouth with my crew but kept it more under wraps around my teachers. I didn't want to spend all of high school in suspension.

I spent all of my time with my two best friends, Cortney and LaRena. We nicknamed ourselves LCM and were basically attached at the hip. Cortney was "the pretty one"—a majorette and part of the pretty-girl crowd that seemed to care more about their clothes and hair than about anything else. They ranged from advanced-placement students to the light-skinned girl with long hair—too stuck-up to talk to the likes of me. Cortney wasn't like that, but she'd hung out with these girls since sixth grade.

LaRena was "the smart one." She was always near the top of our class, a natural at test taking, good in all the subjects. We had very similar families—her parents were divorced, and her dad acted like my mom, pushing her hard all the time. She barely talked to her mom, who worked constantly. We would sit around in one of our bedrooms and bitch about how tired we were of trying to live up to everyone else's expectations. LaRena wanted a man bad. She was always focused on the boys. She was so smart, but she really led with her love.

I, of course, was "the athletic one"—excelling on the team, just getting by in the classroom. I didn't care much about having a boyfriend. I don't think the boys even saw me like that, with my baggy pants and sneakers. And that was fine by me. My mom had badmouthed men so much that I subconsciously didn't want much to do with them. Occasionally I would sneak dudes into my house, not to have sex as my mother suspected but simply to beat them in video games.

Sex was everywhere and nowhere at the same time. My mom never mentioned it, except to warn me—"Men are dogs"—or threaten me—"Just don't get pregnant, Marvelyn. That's all I ask." That wasn't, of course, all she asked, but she said this all the time—instead of actually giving me usable information about sex and how to protect myself from disease and unwanted pregnancy.

School wasn't much better. We had a wellness class where sex ed was buried. The teacher was coach of the all-state men's basketball and track teams; talking about a sensitive subject with a bunch of ignorant teenagers wasn't his expertise. What teenage girl wants to talk about STIs with a man anyway?

I do remember that there was a chapter about STIs and HIV in our health books. HIV was limited to a page, and half of that page was taken up by the picture of a skinny, gay, white man. In other words, not me, not me, not me.

I really never gave HIV, much less STIs, a second thought. Magic Johnson had come out as having AIDS when I was seven years old, a distant memory by then. When it happened I hadn't even been fazed much. I was a Michael Jordan fan, and the *National Enquirer* that my mom read had huge headlines about how many women Magic had slept with. I put two and two together and came up with the conclusion the press was leading me toward.

Though Magic Johnson obviously meant his declaration to serve as a public education campaign against ignorance, I think the black community in Nashville was confused by his announcement. When he said he wasn't going to play basketball anymore because he had AIDS, people reasoned that he had to quit because otherwise he would give it to teammates or opponents, reinforcing the falsehood that HIV and AIDS can be caught like the common cold. People didn't realize that Magic really intended to take care of himself and dedicate his life to talking about the disease. I certainly didn't.

Magic's announcement—really the only obvious opportunity I had to learn about HIV and AIDS—passed by as a blip on the screen of my very dramatic adolescence.

As high school dragged on, things grew worse between my mom and me. I got tired of living her dreams. One day, exhausted and totally not wanting to go to track practice, I snuck home. I knew my mom wouldn't let me just skip it, and that she would hear about it even if I didn't tell her, because she knew the coach. I remember hovering over the toilet and sticking my hand down my throat to make myself throw up. It hurt so badly. I left the vomit in the toilet so that when my mom got home from work she would see it and, I hoped, not get mad at me.

No such luck. She could have cared less. "You better not miss another practice, Marvelyn," she said—her last words to me for about a week.

For my Sweet Sixteenth I got a job at a local Pizza Hut because I wanted money for clothes, shoes, lip gloss, and everything else my mom considered frivolous and wouldn't buy me no matter how much work I did around the house. For one of the first times in my life, I was able to work, get paid, feel capable of something without my mother shouting put-downs from the other room. I was around grease and irritated customers, but none of it bothered me because I was finally truly on my own, doing something for me. I loved the independence.

I also liked it because it gave me a chance to meet guys. When Derek Smith walked in, I was totally sprung. He was cute, sexy, and—the best part of all—twenty-one years old. He went to Tennessee State University and had a house with some hot roommates, two cars, and a Memphis accent.

That summer I lost my virginity to him. I was actually a little late to the game; all of my other friends had started having sex by then. We were very responsible, always using condoms. At the time, I felt no risk because the condoms never broke.

My mom read my diary and found out that I was having sex. I was heated that she put her nose in my business, and she was devastated by the news. She was even more convinced that I wasn't going to, in her words, "amount to anything in life."

This wasn't just her usual tirade. Now she also had this relationship to humiliate me with. "That older guy doesn't want anything but to use you for sex!" she would yell. "He doesn't want anything from a little girl like you who isn't doing anything with her life."

Even though she was cruel, I now realize she was right. That's the worst part. She knew how to bring me down because she hit me where it hurt. Obviously, things might not work out in the long run with Derek, but why couldn't I just enjoy it for the moment? My mom didn't believe in fun. She didn't believe in "the moment."

I ran away repeatedly to be with him, only to be dragged back by one family member or another, or by one of the police officers who knew my mom. To my total annoyance, it seemed like she knew everybody. And while she was ordering me around and telling me daily how little I would amount to, she seemed able to charm everyone else into loving her. I lost my job because of the commotion in my life. Eventually, I lost Derek too. He got another girl pregnant.

That summer, my mom continually threatened to turn me over to state custody. She even took me downtown to the criminal justice center, where, to my relief, they told her to take me home, saying that there were "way worse kids in the world." It seemed like the ultimate statement of how unwanted I was. My mom was so disappointed in who I was that she was willing to hand me over—her own child, her own flesh and blood—to be raised by strangers during the most tender and tumultuous age of life.

I contemplated suicide more than once. It wasn't that I had no joy; I just couldn't imagine how I was ever going to live up to my mom's expectations. When I looked down the long road of my future, I could see only tomorrow—more track practices I didn't want to go to, more verbal abuse from my mom, more close-to-failing grades, more loneliness now that Derek had left me, more of the pretty girls at school posing in the hallways, giving me dirty looks because I wasn't like them.

One afternoon I took the gun from my mom's bedside table, stood in front of the mirror, and held it up to my head. I don't think I intended to shoot. I still had a lot to live for—my nieces and nephews, my daddy, the slight possibility that I might one day get out of Nashville and making something of myself. But I did want to test how serious I was, to feel the cold metal of the gun on my hot temple, to *choose* to live. It wasn't the last time I would test myself in this way. It wasn't the last time I would be faced with a moment where death was just one false move away.

Eventually my older half-sister, Tab, realized how bad it had gotten and provided me with an escape—she invited me to live with her, her boyfriend, and her three kids. Even though the house was packed and I had little privacy, the chaos there was way more comforting than my own house had ever been. I loved those kids—Diamond, Jamiya, and Jamirius—and they loved me back. It couldn't replace a mother's love, but it felt pretty good. I was sixteen and I would never live in my mother's house again.

Things were looking up in general. My friend Shatoya and her mom threw me a huge seventeenth-birthday party. My mom had stopped giving parties for me by the time I was four, so it was a big surprise. "You deserve this, girl!" Shatoya's mom yelled as she hung up a huge banner. I immediately felt like crying. That party was the bomb. It felt like the entire high school was there, and my social status went through the roof.

By the time summer rolled around, I was the closest to happy I had ever been. I had a new job, working forty hours a week at a Subway restaurant. I found genuine love with a boy named James. I finally felt free, and like I might even matter in the world. My mom's voice whispered in my head, but for the most part I was able to drown it out with the wonderful noise of my new life.

When Tab realized that I was having sex with James, she insisted on taking me to Planned Parenthood. She loved her kids, but wanted me to make a conscious choice about when I ended up strapped down with my own kids.

The people at Planned Parenthood gave me a Pap test and a Depo-Provera shot for birth control. I figured that getting the test meant I was free and clear of all STIs, not knowing that a Pap test only indicates whether you have cervical cancer or abnormal cells that might indicate cancer. I told James and we started having sex without condoms. With fear of pregnancy out of the way, we figured we had nothing else to worry about. HIV never figured in my mind.

But the Depo shot made my period painful and irregular. I would bleed for weeks at a time. I hated feeling that way, especially with how involved in sports I was. I never went back for the shot again. But, of course, James and I continued to have unprotected sex. We had gotten used to it, and by then, condoms were just not on our adolescent, ignorant minds.

Money was becoming a problem for me. I'd quit my Subway job so I could focus on sports, my only hope of getting into college since my grades continued to be poor. I didn't like to ask Tab for money, so my boyfriend pretty much gave me what I needed. That made everything even more complicated. I had become exactly what my mom warned me against being—a woman dependent on a man.

I could tell that James was looking around, and it made me crazy. I became one of those desperate girls I despised—totally hooked on keeping my man at any cost. I decided the only way to keep him would be to fake a pregnancy. When I told him, he immediately started to act right again. He was committed to staying in the relationship and making it work for "James junior." I would call him when he got off work and ask him to come by and bring some food—"for me and the baby."

I hated the feeling of lying, so I decided to really try to get pregnant. I was seeing what a wonderful husband and father James would be, and I wanted it for real. But of course, God works in mysterious ways. No matter how much I tried, I couldn't get pregnant. Friends seemed to get pregnant by simply looking at a guy, and here I was desperate to do it and totally unable.

I needed an escape from the lie, so I told him that I had decided to get an abortion in order to pursue my basketball career. Plus, I reasoned with him, we didn't have enough money to raise a baby. He couldn't really argue with any of that. Excited as he'd been to be a father, the relief on his face was obvious. I read up on abortion so I could complain about the correct symptoms afterward. James had sympathy for me, but without the potential of a baby to keep us connected, we had nothing.

One day shortly after Christmas, Tab, her kids, and I went to the mall. All of a sudden, I spotted James strolling past some stores, holding hands with another girl. My heart was crushed. I immediately broke down in angry tears in the middle of the mall on one of the busiest shopping days of the year. People I knew kept walking by and asking if I was okay. It was humiliating.

That incident plunged me back into suicidal thoughts and self-hate. It felt like nothing was in my heart. I quit playing basketball a couple of weeks later, a week before the play-offs. Everyone thought I quit because we were set to play the best team in our district and I was scared. That wasn't even the issue. I just felt useless, as if I wasn't worth a damn. My mom treated me like I was invisible, my dad wasn't around, and I couldn't keep a boyfriend. How was I going to lead my team in the play-offs? Besides, I wasn't a leader, I was a follower.

My suicidal thoughts this time were pushed away by my love for my sister's kids and something my mother's sister, Aunt Beverly, told me: "Those who commit suicide go to hell." I just clung to God and asked for a sign.

I thought I found it in one of Cortney's boyfriend's friends, a handsome guy from another school. Unfortunately, that situation turned out to be complicated. He had a girlfriend, but I was way too desperate to let that stop me.

I moved right on in. I did whatever he wanted me to do. If he wanted some clothes, damn it, I would get him clothes. Shoes, money, sex—whatever he wanted, he got it. I was working just for him. I thought the hard work paid off when he eventually broke up with his girlfriend and started focusing on me. Whenever I would feel his gaze stray, I would buy him a new thing—a pair of sneakers, a watch, a shirt. I would do anything to make the relationship work. The saddest part is that he never even considered himself my boyfriend.

I had seen the most desperate, man-dependent side of myself—just the side my mom had warned me about. But instead of hating that, I fell into it even more. I had tasted how good it felt to be adored by a man. I craved that feeling like my daddy craved crack.

Me, dateless at my senior prom!

chapter three

My senior year was basically one soap opera episode after another. One plotline was my obsession with finding and keeping a man—making me go to all kinds of lengths, from giving up on sports to spending every penny I earned on clothes and jewelry for *him* (not once thinking about Marvelyn). I was still convinced that my only chance of going to college would be through an athletic scholarship, but I had an "I don't care" attitude. I was into living for the moment, never for the future.

When I look back on that time, it makes me sad that I thought so little about my own life. It's hard to pinpoint just why that was. I was certainly distracted as hell, working so damn hard to find and keep a man. Plus, girls were getting pregnant all around me, and they seemed happy, sometimes even relieved. I envied that their boyfriends suddenly paid attention to them and bought them nice things. It was clear that the best way to get a guy—at least for a little while—to appreciate you was to get pregnant with his baby. Now, once you had that baby, well, that was a different story. But then you had a baby, a little person who unconditionally loved you and depended on you. As scary as that sounded, it also sounded really comforting.

I also thought so little about my future because I bought into other people's perception that I wasn't going to have much of one. LaRena was the smart one and Cortney was the one with all the family support and I was the athletic one who didn't have either. That left me just one chance of making it in life: an athletic scholarship, and after that, some sort of professional league. I didn't research the statistics on how unlikely that was, but I didn't really have to. I knew girls who had gotten athletic scholarships, but so far none of them were playing in the WNBA. Hell, I didn't even want to play in the WNBA.

The other constant drama during my senior year was the ongoing feud between Cortney, LaRena, and me and the "pretty girl" clique.

Cortney had been friends with them since she was a child. They were all majorettes together, which at my high school was the big activity for all the popular girlie girls. You could easily spot them. They always looked classy, hair flat-ironed to perfection. You'd rarely see them in tennis shoes; they wore heels almost every day. And they didn't repeat an outfit. Later I would learn that they shopped at discount stores such as T. J. Maxx and Marshalls, but at the time I figured they just pranced up and down the shops at the bougie mall in town. Looks can be deceiving.

The other thing that every popular girl at my school desperately wanted was to be nominated for Miss Whites Creek—the equivalent of homecoming queen. Being into sports, I had no chance in hell and liked it that way. Cortney, on the other hand, had a shot at it. She was a natural beauty, requiring very little effort to look the best. Cortney wore her real hair, no makeup, and stylish clothes—clean and cute. And she got along with everybody.

As we sat in class one afternoon, the principal got on the loudspeaker and slowly read off the names of all the school's superlative nominees. I rolled my eyes and looked around as the rest of the girls in the class held their breath. They acted like it was the

Oscars. When the list had been called and none of the girls' names were announced, one of them, I'll call her Shawn, burst into tears. I laughed my ass off as I accepted the "Most Athletic" title with pride. I admit it. I laughed right in her face. It was nice to see the pretty girls cry and suffer a little.

I liked Cortney, and I respected her choice to be friends with these girls, but I never liked them or respected them. They lived in a fantasy world where nothing mattered as much as their Coach bags or their knockoff Gucci sunglasses. I saw all of them as stuck-up. I figured there were more important things to think about in life, and I had bigger issues to face.

Shawn, who was crying, and a clique of her fake "friends" huddled together and glared at me as I laughed. It was as if their mascara-covered, colored-contact eyes screamed at me, "Watch your back, bitch!" That made me laugh even harder. After all, if they were going to throw down, they would have to break a nail, and I knew how much they would hate that.

I would be surprised. The next afternoon, in that same class, Shawn marched in, glanced at me, and then sat down at her desk, obviously gearing up to talk some shit. I just looked her up and down, calm as a cucumber, and said in my head, "I lift weights on a daily basis, you baton-twirling bitch. What the hell are you thinking?" LaRena, who sat next to me, knew it was about to go down.

"*You lying bitch!*" Shawn screamed. My desk went sliding across the room after I pushed it away, and my fist went straight for her mouth. I grabbed her arm, then proceeded to beat the hell out of her. The teacher couldn't do anything but run out the room and down to the principal's office, screaming for a security guard.

The captain of the boys' basketball team broke up the fight

and screamed at me, "What have you done?" Blood splattered the white concrete floors of the classroom, and the pretty girl's face was anything but. Her last words as she was escorted out of the classroom were "I am going to kill you."

I didn't have a spot on me, though Shawn had managed to pull out the quick weave that my sister Tab had just put in. Poor girl, she'd been trying to hold on for dear life. My only worry was when Tab's schedule would free up so she could retouch my weave.

"You turned into the Incredible Hulk!" one onlooker said. Lots of the guys were throwing jokes. I laughed, but I suddenly remembered my school had zero tolerance for fighting. I knew I would get suspended.

My momma never taught me to stick up for myself. She lived in a fantasy world. She'd advise me to just ignore anyone who was giving me a hard time—sticks and stones and all that. Tab, on the other hand, existed in reality. She had been telling me to beat these girls' asses for a week. She hated hearing my stories about all the petty drama. She thought I needed to put a stop to it.

That incident, of course, did nothing to squash the feud; it only fueled it. Cortney immediately broke with the pretty girls after stating, on the record, that not only did I beat Shawn up but also that she got what she deserved. I realized that Cortney was smarter than I had ever given her credit for. The pretty girls had basically shut her out because they were so mad that she talked to LaRena and me at all. We, on the other hand, would never have shut Cortney out for talking to the pretty girls. We knew that she had a history with them, and we respected that. Ironically, their cattiness separated them forever.

My clique and I were in the principal's office almost every

week, listening to another lame speech about how we were losing our way right as we were about to graduate, how violence wasn't the answer, and how we were not "conducting ourselves as young ladies."

I knew violence wasn't the answer, but it sure as hell felt good. After the humiliation I experienced with my last two boyfriends, feeling so desperate and helpless and unwanted, beating the crap out of that girl actually felt really cleansing. I felt feared and in control, like I could overpower any of those pretty bitches if I needed to protect myself. I knew it wasn't smart, but no one could tell me to stop defending my honor and the honor of my girls. Sometimes it felt like LaRena, Cortney, and my nieces and my nephew were all I had in the world, and I would do anything to protect them.

I would have to do a lot. Shortly after the initial beat-down, one of the pretty girls, with her cousin and her sorority buddy, Sha, jumped LaRena at the gas station, leaving her body shaken and beaten on the pavement. It was low, attacking her when she didn't expect it, taking advantage of the fact that none of the rest of us were around.

That whole senior year we were into it with them constantly. When Cortney was crowned prom queen, it heightened the tension. Fighting and defending my girls became yet another distraction from the impending graduation. I had no idea what I was going to do to keep from being suspended and kept from graduating. LaRena, Cortney, and I decided—for the time being—to let things die down like everything was peaches and cream. We thought we'd get them later.

I had gone to my senior prom by myself, but I dreamed that the guy I was hooking up with at the time, who went to a different high school, would invite me to his. He was one of those teddy bear football players, not fat but stocky. He was a pretty boy. Plaids and polos and Tims were his regular, no white T-shirt and jeans for this guy. I saw him as my last chance to have a night as a princess, my last night to feel like a guy was worshipping me, my last night to—in truth—convince myself that I was wanted.

Plus, I'd given people at my high school, including his ex-girlfriend, the impression that we were together. The truth was that we were just *sleeping* together, a pattern I'd gotten pretty used to by that time. He would never have described me as his girlfriend. He probably would have used the phrase "more than friends" to describe what was going on between us. I didn't want my lie to be exposed, and it would seem pretty suspicious if I didn't even go to prom with my own boyfriend.

I begged him to take me. At first he said he wasn't sure, that he'd have to think about it, but when he called back a week later and told me that he'd take me as long as I paid for it, I didn't skip a beat before assuring him: "Yes, absolutely."

Once I'd hung up, I realized what a mess I'd gotten myself into. My mom didn't believe in proms. She thought they were silly and just a distraction from the important work of becoming some-body worth respecting. She sure as hell wasn't going to give me any money. I was too embarrassed to tell Tab that I had to pay for the entire prom, so I wasn't going to ask her for anything (plus she had three mouths to feed). I knew my grandmomma didn't have much money because she'd recently retired from her job as a secretary. I had no idea how to get ahold of my dad.

So I started working overtime at my new job making shakes and serving at the Steak 'n Shake. As I was making people sandwiches and putting up with their complaints, I would fantasize about how fly I was going to look, how all the eyes in the room would turn to rest on me and "my man" as we walked into the ballroom. I even bought myself a few little gifts—shoes and jewelry—and told other people that he had bought them for me. I could tell that my sister was suspicious about me working so many hours. She must have thought I was on drugs or something. Little did she know I was just addicted to my own fantasies of love and attention.

Funny thing is, that prom actually lived up to my expectations... at least to begin with. I was the princess of the party in my gold, two-piece dress. I found it at a bridal shop in an old mall in Nashville. When I walked out of the dressing room, Tab said, "That's it!" She had never been to her prom, so she was living vicariously through me. She even picked up the four-hundred-dollar tab.

The night was magical. We danced, we drank fruit punch, we mingled with his friends, who clearly didn't think I was his girlfriend but were at least under the impression that he thought I was fly enough to bring to his prom. It always helped to be from a different high school too—it lent you that air of mystery.

Anyway, the night was unfolding beautifully. We left the dance and headed out to Red Lobster. One of his friends decided to show off and rock his car from side to side. He rocked so damn hard he ran right into the car that "my man" and I were in, a rental.

We had to call his mom and dad to come pick us up and take us home. His mom said, "I bet your night was fun until he hit you!" Little did she know that we'd had plans to go off and spend the night at a hotel—more money down the drain, another dream ruined.

The night was over, my fantasy finished almost as soon as it had begun. After that, he and I talked from time to time, but our relationship didn't last long. He found out that a three-year-old child that he'd thought was his was in fact some other dude's. It immediately set off the player in him, and he was ready to be wild again. I didn't fit into the equation.

A week before the prom, I had turned eighteen. My mom gave me a real surprise. Not a surprise party, never that. Naw, she decided that the day I turned eighteen was the appropriate time to finally move all of my stuff out of my room (where, admittedly, I hadn't lived in a few years) and move my little sister right in. She turned Mone't's previous bedroom into a decked-out office. Don't ask me what my little sister needed with an office at fifteen years of age. Don't ask me why my mom felt the need to physically remove any trace of me from the house the second I became an adult. I felt, once again, rejected and devalued next to my perfect sister.

I graduated on May 21, 2002. It was a beautiful sunny day. My mind was practically blank with joy. I had done it. Despite all the fighting and the almost failing grades, I had managed to graduate. I just hoped that when my name was announced, people would cheer.

My mom, Mone't, Tab and the kids, and even my dad helped fill Tennessee State University's Gentry Center.

"Marvelyn Brown." Once my name was called to receive the diploma, the only thing that I could think of was walking in a straight line and not tripping from the painful shoes that I was wearing. The crowd, indeed, cheered. With that diploma in my hand, I smiled from ear to ear . . . not because I had graduated, really. I didn't think it was that big of a deal. I smiled because I was free.

Afterward, LaRena, Cortney, and I posed for pictures. We all looked fly, and as usual, we knew it.

When people would ask me, "So, what's next?" I could only smile back at them and shrug. I really had no idea.

LaRena, Cortney, and me (L to R) on graduation day.

chapter four

The summer after my senior year, I felt like a half-changed woman, living day by day. Everyone else was getting all serious, heading off to college, and I suddenly had this freedom; all I wanted to do was act like a kid, be silly, have fun. Things were undoubtedly serious for my best friends. LaRena was pregnant *and* had a full academic scholarship to Tennessee State University. Cortney was headed off to Middle Tennessee State University. They were making moves, but I had nowhere to go.

Except the clubs, that is—I went every chance I got. It was eighteen to party, twenty-one to drink, and, damn, did I party. I didn't drink. I didn't have to. I never had a hard time striking up a conversation with someone, especially a guy. Sometimes I'd be talking to one, just chatting about whatever, and all of a sudden he would remember I was a girl. "Damn, I can't believe I'm telling you all this!" he'd say. "It's like you're one of the dudes!"

Though I didn't have a hard time talking with the guys, I was starting to realize that I didn't necessarily want to be one of them. I knew there was something more to me than my sweats and my sneakers, that after all that fighting with the pretty girls, maybe there actually was a pretty girl underneath my tomboy façade. (I would never be petty like them, mind you, but I could be pretty.)

I started to wish for a little of that "Damn girl, you look good" attention that other girls received, in no small part because of the way they dressed, their hair, their nails. I had never cared about these things before, but without an athletic scholarship in sight, my tomboy style wasn't getting me anywhere anymore. Maybe the pretty girl underneath could take me places?

At first I just experimented. I would be posted up on the wall in the club, just like a dude, but wearing a sexy dress. My auntie always told me that I had pretty skin and shouldn't use makeup, so I never really did that, but I started borrowing Tab's shoes when she wasn't there. She has a Kimora Lee Simmons closet, so I figured if I took a pair, she would never know they were missing.

The first time I borrowed a pair of her peanut butter knee-high boots, I remember almost breaking my ankles. I'd always heard that "beauty is pain," but those shoes were beyond pain. What's beautiful about winding up in a cast? I just wasn't buying it at the time. It took a while for me to figure out the secret—wear the flats and carry the heels until you need to look cute and impress someone. You just can't get the same reaction in the flats; sometimes you gotta retire them.

Tab was largely responsible for my transformation. She saw me coming home from the mall one day holding yet another new box of sneakers, and she said, "You know how many pairs of pumps you could get for that one pair of sneakers you just bought?" I always was a bargain shopper, plus I always listened to my sister. She explained how stilettos enhance a woman's figure. I trusted her opinion because she was all I had in the way of a role model. I could look cute, save money, and get boys too? Come on now, that sounded like a successful plan.

When I found the Air Force 1 high heels, I was in heaven. They were the perfect hybrid of my two selves. I wore those boots out. In fact, when I think about that time, those boots basically sum up the whole experience. I was still Marvelyn—tough, scrappy, with an attitude—but I was also discovering a softer, feminine side.

Without college on the horizon, I started waitressing at the Steak 'n Shake. I loved it there. I was a people person, so tips came easy. All the activities that my mom tortured me with as a child— performing, competing, networking—were finally paying off. I would saunter over to customers and give them my million-dollar smile, bring them free refills, ask them about their day, and before I knew it I had a way-over-20 percent tip waiting for me. Of course waitressing at the Steak 'n Shake was not my momma's vision of what I should be doing, but I didn't care. I was making good money—enough to keep me clubbing and shopping to my heart's content. I enjoyed myself. I felt good at my job.

I lived *only* for the moment—if I was having fun, that's what counted—and I was always taking shortcuts. Of course I wanted more out of life than to be the top server at a whack-ass restaurant in Nashville, Tennessee, but I was not qualified for decent-paying jobs. I thought, "If you serve and you do it right, you make some decent money. Hell, I can serve for the rest of my life."

I wasn't thinking about the plan God had for me down the line, deep fulfillment, making the world a better place, all that. I just wanted to have a good time. My "let me just get a D" men- tality from high school evolved into "let me just get a little bit of money and then go have some fun" mentality in the real world.

In some ways, I felt like I deserved that. My childhood had been rough. My momma had criticized me, and my daddy had repeatedly disappointed me. My friends got to go off to college while I didn't have a single school knocking on my door (and I did think they had to knock on your door, not the other way around).

It seemed like I was entitled to a little entertainment in life.

But the summer became the fall, and I had a rude awakening when Cortney and LaRena deserted me for bigger and better things. While I was asking people "What kind of shake would you like with your burger?" Cortney and LaRena were moving into the dorms. Both text-messaged me to let me know that things were safe in their rooms, but I didn't reply. I was excited and proud for my girls, but I can't say there wasn't some jealousy there too.

As if God heard my inner yearnings, the phone rang one day while I was clearing a table full of half-eaten burgers and sticky milk shakes. My manager called me in. "Marvelyn, you got a call! You can take it in my office."

I dropped the dishes off in the kitchen, went to the manager's office, and closed the door behind me. "Hello?"

"Marvelyn, it's your momma."

"Momma, you shouldn't be callin' me at work."

"Just listen girl, I've got exciting news. This coach at Savannah State called. He was looking at your tapes and he said that if you get your ACT scores up and enroll in junior college next semester, he might be able to get you in to play basketball next year."

My stomach lurched. This was exactly what I'd been waiting for. It was the knock on the door. "What do I have to do?"

"Well, you got to start working out, for one thing. The next ACT test is October twenty-sixth. You should probably take a prep class so you can raise your score. It has to be at least an eighteen."

Now my mind was racing. Work out? Take an ACT prep course? Life was going from silly to serious pretty fast. "But you want this," I reminded myself.

"Cool, cool, Ma. I'll call you later, okay?"

"Okay, Marvelyn. Just don't blow this. It might be your one

chance to make something of your life." As if I needed to be re-minded of that.

I hung up the phone and thought to myself, "Wow, this is it." Then I quickly got back to slinging burgers and fries.

My life went from all-over-the-place fun to a regimented schedule. A day in the life of the finally-getting-serious Marvelyn Brown:

9:00 A.M.	Go to the gym and lift weights or run
10:00 A.M.	Study for the ACT
11:00 A.M.	Go to ACT prep class
2:00 P.M.	Go to Whites Creek and play basketball
5:00 P.M.	Dinner shift at Steak 'n Shake (don't eat any of the food!)
12:00 A.M.	Go home and pass out
Rise and repeat.	

I convinced my old basketball coach to let me practice with the high school team, because I knew that there would be no other way to really get into the pristine condition that would impress this Savannah coach. Though I had quit basketball my senior year, he'd never quit on me. "You have so much potential, Marvelyn. I'm glad to have you back," he said. He was a coach of champions, someone who was used to seeing his athletes win and win big. I was grateful to be back in his presence.

I paid for the ACT prep course and my gym membership out of my own pocket, which motivated me to go to both religiously. The ACT teacher was entertaining, thank God, because otherwise I never would have been able to pay attention. She had a warm personality and was charismatic, even while explaining something as boring as a standardized test.

Honestly, it was a lot of hard work, but it felt good to be back. As I ran around the court each afternoon, I was able to recapture that feeling of being the go-to girl, someone important and celebrated. In high school I would come out of the locker room and people would cheer. I was always the first one to start the layup line. I hadn't been able to admit to myself how much I missed that feeling. I'd messed up my high school years so badly. This was my second chance.

It was funny because I'd never envisioned myself getting a basketball scholarship. I loved playing, but it was track and soccer that I really excelled at. Truth is, I was never good at getting along with a team. The other girls would pick on me because my momma would come in and talk to the coach about my playing time (the kiss of death for team spirit).

I led my freshman team all the way to the city championship and hit the game winning shot, but when it came to varsity, things were different. The older girls could miss twenty thousand times, but if I shot the ball and missed once, I would hear about it. I remember one game where the opponents doubleteamed up on everyone except me, because they were like, "She's not going to shoot the ball." We won that game, despite being the underdog, because I just kept shooting. I'm pretty sure I still hold the record for the most three-pointers in one game. My coach celebrated my success, but the rest of the girls never said so much as a "Good job."

This time it was just about me, myself, and I. This was just about finally doing my thing and pursuing a goal.

Sometimes what you experience in life actually confirms the sadness in your heart. These moments can be strangely comforting, the way it all matches up, the way the world reflects that things are as hard as you have felt them to be, that people are as shady, that existence is as fragile. For me, such a moment came full force in October 2002.

I was in between a workout and a shift at the Steak 'n Shake when I got a call on my cell phone. It was Cortney's ex-boyfriend. When I saw his name on my caller ID, I scrunched up my face. "Why the hell is he calling me? He knows that I'm busy and Cortney has moved on," I thought to myself, or maybe even said out loud (I have a habit of talking to myself sometimes).

There wasn't much small talk. He just came right out with it: "LaRena shot herself. She's in the hospital."

I threw down the phone, as if it were the cell itself that had delivered this unbelievable news. I felt my whole body seize up in protest, and I started screaming and crying—an immediate reaction to what I knew in my heart to be true.

It was old news that LaRena's man had been cheating on her despite the fact that she was eight months pregnant. It was a favorite rumor among the kids who used to go to our high school, and LaRena knew it. Cortney and I had gotten sick of hearing about him and all his infidelities. We wanted LaRena to just deal and move on. We, of course, had no idea that this was how she would "deal."

I had known LaRena since I was ten years old. We'd always been on the same summer track team. When I first met her, she had glasses and braces, but over time she developed into a beautiful girl—half black, half Asian, spiral curls, no makeup. She was a tomboy like me, but a pretty tomboy. She still wore tight pants and tight shirts, but she would have Air Force 1's on too. She was very smart, loving, and careful. She talked in a baby voice. That's how sweet she was.

When she was little, she always had a book in her hand, so I was sure she'd be a nerd. While she continued to do well in school and get those A's, she was also a lot of fun. She, Cortney, and I would hang out constantly—driving around, going out to eat, or shopping.

At our senior prom, LaRena and I were both dateless. We sat there—she was in light pink sequins head to toe, I was in shimmery light purple—waiting for them to announce the prom queen. When we heard Cortney's name, we jumped up and down, screaming at the top of our lungs, "That's our girl! That's our girl!" I remember thinking, "Oh my God, we're going to fall. That, or my dress is about to split." LaRena and I had these dresses on, but the athlete came out in both of us. We were just so happy for Cortney.

Another time, LaRena and I got into trouble in theater class. We were the star characters—Missy and Lutiebelle—in scene one of that year's school play, *Purlie*, written by Ossie Davis. It was all about this preacher who inspired the remaining slaves on a southern plantation to rise to freedom, but I have to admit that we weren't really paying attention. We didn't perform again until the third scene. LaRena and I got so bored backstage that we just started throwing tissue on the stage. When the theater teacher came back and yelled at us, LaRena looked at her with big eyes and said, "We got bored." We both burst out laughing.

She was my best friend.

I couldn't imagine life without her.

I managed to pull it together enough to get a ride to the hospital.

When we finally got there, we rushed into the waiting room and found LaRena's whole family sprawled out on chairs, crying and wailing, comforting one another, some sitting in stone-cold silence and staring forward in shock. I didn't know what to say, what to do, so I just sat beside other high school classmates and closed my eyes, started praying as hard and fast as I could for that baby inside of my best friend. I didn't know anything about her diagnosis, if the baby even had a chance to live, but I did the only thing I knew to do in that kind of a situation. I prayed hard.

My prayers were interrupted when the emergency room doctor came in and went right over to LaRena's dad and told him that, though LaRena wasn't all the way dead, she was as good as a vegetable. She would die soon. He asked if the family and select friends wanted to come in and say good-bye. LaRena looked bad, he warned.

"And the baby? What about the baby?" LaRena's dad asked, desperate for a shred of hope.

"We've put him in our neonatal intensive care unit. We're doing everything we can to save his life, and he actually seems to be in pretty good shape."

You could watch that little shred of hope wash over LaRena's daddy. It was amazing how much this meant in a time of such total devastation. It was almost like LaRena would live after all, in a different form.

Sometimes I wish I never went into that hospital room to say good-bye. LaRena's face was all swollen (she had shot herself in the temple), and the bump from her pregnancy had deflated as if it were a balloon that lost its air. Her momma sat beside her and just kept kissing her and wailing and kissing her some more. I felt like she was mourning a stranger. LaRena didn't look like herself.

I closed my eyes and said to myself, "I forgive you, LaRena, and I am not mad at you now. You are happy. You have gotten what you want."

When I got back out to the waiting room, everyone was quiet. It was like the truth of LaRena's death had produced a silence that would never lift. We all sat there, unspeaking, unmoving, for hours, in a vigil for the baby boy who was our last chance of not losing all of LaRena.

Eventually I would learn the details of that day. LaRena, hearing of her baby's daddy's infidelity for the last time, went over to the hotel where the other woman and her baby daddy were staying. She confronted him there and shot herself square in the head right in front of him. It was a desperate, gruesome way to die.

LaRena's supporters vilified the baby's daddy, but I didn't feel like that was entirely fair. Sure, he had caused LaRena a tremendous amount of pain. Sure, he had acted irresponsibly and unfaithfully. But LaRena pulled that trigger. She made that choice. I had empathy for what he must have been going through, dealing with the guilt of playing such a huge role in the death of a lovely young woman. It wasn't that I liked him any more than anyone else did, but I just didn't think that he should be held totally responsible for LaRena's passing.

In the days following her death, I thought a lot about the choice she made and what followed. As I've confessed, I had considered suicide myself at various moments throughout my life, even felt the cold metal of the gun against my temple. But watching the way LaRena's death was perceived by others, was manipulated by near strangers into their own excuse to get the spotlight, sickened me. I realized that when you died, you no longer had control over the way your life was commemorated. You didn't get to deliver the final message. Other people did, and damn if they often got it totally wrong.

Cortney flat out could not deal with the whole thing. As soon as she heard about LaRena's death, she turned off her cell and wouldn't accept calls. I didn't blame her. It got tiresome fast to field calls from people who hadn't even cared about LaRena when she was alive—even the pretty girls—asking how she died. People would ask me, "How you doing, girl?" and I knew they just wanted a piece of the drama. People crave drama at the most inappropriate times.

That's why when it came to making a RIP shirt, I wanted mine to be different from everyone else's. I took a long-sleeved T-shirt and put her prom picture on it. Cortney and I decided we'd use that picture because only eight people had that picture, those who were truly closest to LaRena. Everyone else put her senior picture on their shirts, walking around like "Oh, my God, I knew her." Not us. We had LaRena in her pink prom dress. I also wrote "Rena B," because that was my nickname for her. I took a pair of LEI pants and put "RIP LaRena" down one side and "1984–2002" on my butt. I wore that outfit every single day until the funeral.

The day of the funeral was strange for me. I had been crying for days on end, mourning the loss of my best friend, my "smart one," my track partner, my girl for life. But when I arrived at the funeral home, it was like the rest of the world rushed in. People were everywhere, including the pretty girls with whom we'd been feuding the whole last year. LaRena's baby's daddy was not there, nor was her unnamed son, who was still in intensive care.

LaRena was buried in her pink prom dress. When Cortney saw her body, she just stood next to the casket, paralyzed. When she gave a eulogy for LaRena, she sobbed so hard through the entire thing that you couldn't understand a word she was saying. She'd lost another best friend as a young girl, so the pain was too familiar.

At LaRena's funeral, the class president got up and apologized for her friend, one of the pretty girls, for all the fighting. People showered LaRena's casket with flowers and declarations of love. Everyone seemed to suddenly have something they needed Rena B to know before she left this world.

I tried to make myself cry in that moment. I tried to make myself feel some of what they were feeling. But mostly I just felt mad. I looked at LaRena as I passed by her casket and said, "Oh, Rena, I wanted you to be so much stronger, so much tougher." I knew she was going on to a better place, a grand place, and for this I was thankful. I even thought to myself, "This world is crazy. I guess when I get up there I'll already have Rena B waiting for me."

Looking around at all the mourners, I felt sick, as if they thought this whole thing had been some kind of joke, some play in which they were starring. LaRena was dead. She couldn't see all these wonderful flowers, all these apologies and memories. Where were the flowers when she was alive? Where were the apologies then?

Funerals are never for the dead. They're for the living. But La-

Rena's death made me understand that even more clearly—when you die, you have no control over the way you are remembered. Your death—especially when it's a suicide—is never the message you wanted to send. People write their own messages. They write their own stories of your life. I knew, then and there, that I didn't want that. I wanted to be the author of my own life story.

My mother, hell-raiser and feminist.

chapter five

LaRena died on October 20, 2002. Her funeral was on October 25. The ACTs were on October 26.

I planned on postponing until December, the next time that the test was being given, but everyone around me said, "Don't dwell." Dwell? It had been one day since I buried my best friend. After the hoopla of the funeral was over, it seemed, people had no capacity to remember the dead, to honor them with authentic grief. My grief didn't fit into the allotted slot of time that the memorial hall had given for "remembering LaRena."

But even my momma thought that I should go through with the test: "You have to move on, Marvelyn," she said. "LaRena would want you to live your life." I supposed she was right. It was LaRena who had always focused on school and been so dedicated to her work. Maybe I would somehow be filled with her spirit on the day of the test.

I arrived at the testing site early, my RIP outfit newly washed and ironed, a few number-two pencils clutched in my hand.

I needed an 18 in order to get into Savannah State. That number rattled around in my head for the next few weeks like some kind of curse. I had no feeling for how the test went. It was one big blur. I had never been studious, and though I'd tried really hard to prepare for the test, I knew I was better working out in the gym than working out a math problem any day. But maybe, with the help of LaRena's spirit . . .

I kept myself busy so I could override my nerves—playing basketball, working hard, working out. Cortney and I had lost touch a bit after she went up to school, but bonded by LaRena's death, we were talking all the time again. She would tell me about college life, and I would fantasize. It motivated me even more on the court and in the weight room.

Just a week after LaRena's death another dude in town died by suicide, and everyone was running around saying, "He pulled a LaRena." I didn't know it at the time, but suicides often take place in clusters, something that people who work in suicide prevention try to keep on the low so that it doesn't become a fad. People in Nashville still, to this day, call a suicide "a LaRena"—showing just how memorable her death was.

I had to get my mind off of the ACTs and LaRena's death, but the clubs held no appeal for me. So I started heading up to Cortney's school, Middle Tennessee State University (MTSU), whenever I got the chance. If I wasn't working, or on my way to work, you'd better believe I was up in her dorm room—just a forty-five-minute drive from Nashville. In fact, I spent so much time there that some people started to think that I was enrolled (little did they know I was praying over my ACT score nightly). We partied a lot after LaRena passed, our way of forgetting and also our first introduction to college life (mine, of course, without the stress of classes).

Being around Cortney and all of her college friends was supposed to be a vacation from the giant, blinking 18 in my head, but instead it got me thinking a lot about my own future. As much as

I liked partying with them, the lifestyle, the atmosphere, it bothered me that I wasn't actually a student there. It wasn't as if I was ashamed, but it did eat away at me a little bit. I wanted that green light to go to Savannah. I wanted to make something of myself on my own terms.

When I got the envelope, it was as if my heart felt torn in two, not knowing whether to sink or soar. It would be either the beginning of the rest of my college life or the end of my dream of having one. Before tearing the envelope open, I prayed one last time. You know how you ask God for "just this one little thing"? It was one of those moments—a "just this one little thing" prayer.

But "just this one little thing"—as is so often the case—wasn't. I got a 16. The wind was knocked right out of me. All that work, all that studying, dealing with the sudden and violent death of my best friend—and I was two points short. I'd done the one thing my momma had always warned against: I'd failed.

I didn't want to tell her, but I knew I had to. As one of our rare conversations on the phone drew to a close, I spit it out, "I made a sixteen."

Silence.

Dial tone.

Back to square one. Some of the girls at Steak 'n Shake went to school at Volunteer Community College, a short drive down Gallatin Road from Nashville. When my momma told me I would lose the spot on her health insurance if I didn't enroll in school for the next semester, I decided that Volunteer was as good an option as any. I was back to going with the flow, not making much of my future, living for the moment. All of the determination and work ethic I'd built up before LaRena's death and the ACTs had drained out of my mind.

That first semester I had to take all remedial classes because of my ACT scores. It made me feel so dumb. I remember sitting in basic math and, in my head, correcting the professor. I felt that school was a waste of time. I was always smart but I never applied myself. When I found out I could pass with a C or a D in high school, I never understood why I would do that extra work. What did it matter whether I was at the top or bottom of my class as long as I walked across that stage? I'm sure I finished at the bottom, grade-wise, but I know I wasn't dumb. That approach didn't serve me when it came to the tests. I hated taking them. It wasn't my thing. I would start doing little designs as I filled in the bubbles. That was more entertaining than actually trying to figure out every single question.

At this time I was still going up to MTSU to party with Cortney a lot, and sometimes she would come down and visit LaRena's family and the baby—who was thriving even without his momma. It made Cortney feel connected to see them, but I didn't want anything to do with it. That may sound harsh, but seeing him made it hard to cope—like he was the ghost of my best friend, all that remained of her spirit and smile.

And besides that, I was always worried that something would happen to the little guy. One time Cortney and I went and picked him up and we were driving him around, and I felt like I had the president of the United States in the car. If we got into a wreck, he would be gone. He was all any of us had left of her, and I was too afraid to be responsible for his life.

At one point Cortney called me out on it: "Why do you act like you're scared of him?"

"Because I am, Cort," I told her. "He's all we got left. I can't lose any more."

Once again, I was in shortcut mentality. I figured I could keep going to community college for two years, get my associate's degree, and I'd be finished. I even planned on going in the summer so that I could play catch-up and possibly reduce the time. Anything to be up and out of school faster.

But I needed a damn car and a license. It was too stressful to always be figuring out how to get rides from people. No one had ever taken the time to teach me how to drive, but I'd observed people doing it. I was like that—if there was something I wanted to learn, I would watch others and try to figure out just how they did it. I learned all the sports that way, especially soccer. I would flip through the channels until I found Mia Hamm and then study how she played. As a result, the first time I stepped on a soccer field, I wasn't a novice. When I got determined about learning something, there was nothing that could stop me.

Driving was no different. Other people take classes or have their daddies take them out in an empty parking lot. I simply asked a coworker to drive me down to the test site and then hand me her keys. When I got behind the wheel of her car and I took my driver's test, it was the first time I'd been behind the wheel of a car at all. And I passed.

Then all I needed were the wheels. My mom and I had been on better terms since I'd been going to school. Though I knew she wasn't impressed that it was a community college, I think she saw it as a step in the right direction, like I was getting my life together.

She had a friend who was getting rid of an old, wrecked-up car, and my mom not only helped me pay for it but also explained how car insurance worked and offered to put it in her name as long as I would pay her back. That suited me just fine. I wanted to grow up, but being completely independent scared the hell out of me.

Having a car was amazing. Of course it was embarrassing that it looked like it was sneezing in the back—as one of my close friends said—but at least I was rollin', and that made me too happy to care. The car worked well, despite looking like a real hoopty. I finally had my freedom.

Except, of course, as soon as my mom gave a gift, she started holding it over my head. She had no tolerance for my lateness with insurance payments, even though I was waitressing *and* going to school. We started fighting again, with just as much heat as before. Our honeymoon was over before it even started.

I was spending more and more time up at MTSU with Cortney. It felt like my little vacation spot—away from my mom and that drama, away from the restaurant and that drama, and away from school and that bore. In truth, we barely escaped the drama at MTSU. One time I was up there with Cortney and she said we'd drop by one of her friend's rooms on our way to a party. To my shock, we walked into a dorm room to find Sha getting ready for the night. She welcomed me with open arms, but I was hesitant to respond in kind. I pulled Cortney aside and said, "Isn't that the girl who ganged LaRena?"

Cortney went all grown-up on me. "Marvelyn, it's passed. Let's let it go. LaRena is gone and it's time for peace." I had to respect her wisdom. I'm grateful I did. Sha and I became good friends over the years.

I wasn't discovering much about my brain up at MTSU, but I sure as hell was discovering something about my body. I remember one of the first times I went up there and partied with Cortney and our two new friends Kendria and Sha. I put on the Off-Broadway version of the Manolo Blahnik Tim boots with the heels, the tight, bubblegum jeans, and this random pink halter top. I was just walking around with them and all of the guys were hollering at me! They weren't hollering at Cortney, who had always been the pretty one, at Sha, who had on the real Manolos, or at Kendria, the typical light-skinned bombshell.

"I guess that big old booty of yours is getting you some attention, finally," Cortney said, playfully jealous. It was as if someone had suddenly turned the light on. I realized how powerful it felt to be pretty, and that, to my total surprise, I was. I ain't never been ugly, but I never really thought about my looks that much; they were always an afterthought. Now they became my calling card in this new phase of my life. My transition came at the perfect time. The Manolo Blahnik knockoff Tims and the Air Force 1 heels came out. There were all kinds of hybrid tomboy pretty-girl styles for me to rock in. That Jay-Z song featuring Beyoncé—"Bonnie & Clyde "—was bumpin' everywhere. Life was good.

Before long, Sha, Cortney, and I all had someone on the basketball team. It wasn't as if he was my boyfriend, but he was cool to hang out with, and more often than not, we would mess around at the end of a night out. We used protection. We all used condoms with anyone who wasn't our boyfriend. That was sort of the unwritten rule. You obviously didn't want to get pregnant from some random dude, but otherwise there seemed to be little to worry about.

Which was probably just the attitude that made me lose my job. We would party late at night, and then, as Sha, Cortney, and Kendria slept in, recovering, I would have to get my ass up and drive the forty-five minutes back to Nashville to make the morning shift at Steak 'n Shake. One night we stayed up until six in the morning, drinking and talking, and I figured I should just drive home right then, take a little nap in Nashville, and then get to work by nine.

Except, of course, I never woke from that little nap and slept through my whole shift. I knew they were going to fire me, so I marched right in there and quit. I never liked to give anyone in authority the satisfaction of rejecting me. I would reject my own damn self.

But for all that bravado, I now had issues. Namely, I had no money. I would get it from random places—my grandmother, Tab, who was a cosmetologist and made a handsome salary. I asked my mom for help, but of course she said no. I would hear that ugly word and just hang up, as usual. It made the memory of our little honeymoon period sting even more, like she had lured me back in just because she was going to eventually get me where it hurt. I was obviously dependent and broke, just like she'd warned me I would be if I didn't get serious about life.

I didn't want to address any of that, so I just avoided her like the plague. I would mostly shuffle back and forth between my sister's house, my grandmomma's house, and my aunt's house. In fact, in my Aunt Wanda I was finding a real friend. She was my mom's brother's wife—far enough away from my mom to see me for my own person and not swallow whole my mom's version of me. Instead, I was her favorite. She was only thirty-two at the time, so we weren't that far apart in age, and we had a lot to talk about and bond over. She ran her own day care, which I respected a lot; to run your own business, help kids, and take care of your own bills all seemed really admirable to me. She did it all on her own terms. Plus, unlike my mom, she actually seemed to have fun. It made me realize that you didn't have to be deadly serious just because you were an "adult."

When I would stay over at her house, I usually shared a bed with one of my two little cousins. They also became my little rocks. Throughout most of my teenage years it seemed as if I was always sharing a bed with one of my cousins, nieces, or nephew. These little souls were often my saviors. They loved me unconditionally. They didn't know anything about the job I lost, the partying I was doing, the classes I was flunking. They just took me for who I was, in that moment.

The close relationships I had with them also helped prevent me from getting pregnant like other girls I knew. It wasn't that I didn't want kids just like them; I did. But my relationships with them filled the void that other girls filled by getting pregnant. I already had my unconditional love, my sweet little faces, those tiny little feet to dress up in the latest miniature Air Jordans. So it turned out that these little kids saved me in all kinds of ways. To this day I am so grateful to them.

I started working for my aunt's day care, which required that I be ready at 6 A.M. It was rough, but at least I worked for someone I respected, and more often than not, I could just roll down the stairs to get to work. She never gave me shit about how late I'd been out the night before, as long as I was there at six, ready to help out with the kids.

I had lost all desire to get to classes, much less study in order to pass them. By April it was clear I wasn't going to finish the semester with even one credit under my belt. I was disappointed, but it was hard to care when I wasn't even that invested in the first place. I wasn't working at the Steak 'n Shake anymore, and that was where all my community-college friends were. Without that culture around me, it was impossible to motivate myself.

While I didn't care that much that I was flunking out—besides the obvious wrath I would get from my mom when she finally heard—I did care that Cortney was flunking out. All the partying we had done had gotten the best of her too. It wasn't a big deal for me to do badly in school. I'd never really done all that well. But Cortney had always applied herself. School was important to her.

I didn't dwell on my guilt too long. Winter was becoming spring and I was feeling like starting over. My aunt's house was mostly drama-free, a welcome change. I wasn't in school, so there wasn't the daily drag of making myself do something I actually hated. Working with kids was a breath of fresh air. I started to feel optimistic about life.

Sha and Kendria, the MTSU Girls.

chapter six

One beautiful spring day I headed to the park with my aunt to watch my little cousin play ball. The sun was shining and that damp Tennessee heat was just starting to creep in. Things were getting sultry.

When I got to the park, however, it wasn't my cousin I was focused on. There was this guy standing off to the side of the court with a group of friends. He was one of those light-skinned dudes, nice, neat braids, pretty eyes, one of those guys who can wear a loose shirt and you can still tell that he's muscular. He was on crutches, and I remember thinking to myself, "How can someone look so good on crutches? All the guys in the park today, and I can't stop looking at him?"

I wasn't watching for any particular end. In fact, I concluded right away that he was out of my league. Eye candy, nothing more. Though I had gained more confidence in my looks as I started paying attention to them more, I still didn't think of myself as the kind of girl who could snag this kind of guy. He seemed older. He had a certain glow about him.

I had a good time just looking . . . until I noticed that he was looking back. "Oh shit," I thought, "it's time to make a move." But of course I was too chicken, unused to approaching guys in a setting like this. Sidling up to someone at a club or a college party was different. Everyone knew they were there to hook up and get down. That greatest of all social lubricants, alcohol, was involved. The park that day felt way too sober.

A friend of mine—who was even younger than me—went over and told him that I thought he was cute; could we have been more juvenile? He took one look and then, smiling a very sexy smile, hollered, "Come over here. I'm going to give you a chance to redeem yourself."

I felt like I was back in third grade and my teacher had called me out during modeling class: "Posture, Marvelyn! Posture!" I smiled back sheepishly, stood up straight, and sauntered over. Sauntering, mind you, was no easy task, as I was up on some bleachers. As I was making my way down, I just kept repeating to myself, "Hit those steps, Marvelyn. Don't trip. Look fine." Miraculously, I made it over to Prince Charming without getting myself into *America's Funniest Home Videos* along the way.

"You have beautiful eyes, girl. Good thing you finally came closer so I could take a look at 'em," he said. He had me at "you."

We exchanged phone numbers, and I hate to admit, I threw all the rules out the window and called him that very afternoon. I was too curious to find out if he'd given me the real digits. He had. And damn if he wasn't still charming on the phone. He invited me to his house.

I told my aunt that I was headed to my grandmomma's house and that I'd see her after church the next morning. I was a grown-ass woman, so it wasn't like they checked up on me, but I knew I'd get an earful from my aunt about playing hard to get if I told her what I was doing. It was a bold move, but this guy made me feel that way—like throwing everything else out the window when he turned on that charm.

When I arrived at his apartment, I discovered even more reasons to fall for this guy. Not only did he have his own place, but it was a man's place, decorated to the nines. He was twenty-four (I was still nineteen), had a job and his own car, and paid his own rent. He was every southern girl's dream on paper—and in the flesh. He had me fantasizing from the start.

But we didn't go out like that right away. Instead, we got into this long talk about sports. He had been a big athlete at his high school back in Chattanooga, Tennessee, going to the state basketball championship almost every year. He still had all of his awards and newspaper clippings in albums. We looked through them and rehashed our glory days. He was obviously impressed at what a star athlete I had been too.

"You played sports? You're too cute to be on the court," he said, "unless you were a cheerleader."

"I was a tomboy in high school. I played every sport but tennis and golf. I was voted 'Most Athletic,' and I was a starting freshman in basketball *and* soccer."

"Soccer?" he asked.

"Yeah, I been sacking balls way before you," I said playfully.

"You really earned that 'Most Athletic,' then."

We talked and talked, and it seemed like we could have talked forever, about anything. It didn't matter whether it was serious—our families, our histories—or dumb crap—cartoons, the latest jam—we always seemed to be on the same wavelength. We just had that kind of vibe, as if we'd known each other for years.

It was clear that he was into me, but I was also aware that I was younger, probably seemed pretty vulnerable to him. Though I was thinking, "This must be what love feels like!" I was trying to give off the impression that it was just cool—him, me, hanging out, nothing crazy. Inside, my whole body was excited about the idea of finally meeting someone who was not only dope but who seemed to understand me from day one.

I spent the night, against my better judgment, although I swear we just fell asleep talking. It was as if neither of us wanted the night to end.

In the morning he made me some breakfast—eggs, bacon, the whole nine. I sat across from him, but my head was really in the clouds. My life had transformed in just one day. I had gone from feeling happy but pretty lonely, hungering for that one guy who would really take the time to get to know me and understand me, to totally satiated. When we parted ways, he gave me a respectful kiss on the cheek and a long, comforting hug. And then that was that. I went back to my aunt's house and started my life all over again.

I've always been the type who falls for guys fast and easy. One moment I'm just sizing them up in the park, the next moment I'm naming our babies. I don't like to admit that. I know about the art of seduction and playing hard to get and all that, but when you really fall, it's hard to restrain your impulses to leap all the way.

I started going over to my Prince Charming's house all the time. I would skip work, or go straight from work and bring him some pizza. I loved spending the night in his bed, feeling his warm body against me. After years of moving around, never feeling truly loved and accepted, being with him was like finally having a home.

As you might imagine, we got down pretty fast after meeting. We always used a condom though, and I figured that if you really loved somebody, there was no real harm in being sexually active. I wouldn't have called him my boyfriend, but I felt like we were moving in that direction.

I kept our relationship pretty much on the down low. I like my business to be private until I know it's going to stick. I'll wait

a week, sometimes a month, before I mention a man I'm seeing—
even to Cortney.

My aunt figured out what was going down pretty early on and sent
me a loud and clear warning: "He's out of your league, Marvelyn."

"What's that supposed to mean?"

"Not that you're not worthy of him, just that he's a player. He's
on another level that you're not ready for."

"I'm ready," I told her, making a cocky face.

"Play with the big boys and you're going to get burned," she
said, shaking her head. "That's all I'm sayin'."

I appreciated her concern, but things felt under control.

How wrong I was.

One night at work, bored as hell, I took one of the remaining
cheese pies (we were allowed to take them home at closing) and
put the first letter of my Prince Charming's name in pepperoni on
the top, then baked it just right.

I called him on my cell on my way to his house and told him
I was coming over and that I had a surprise for him. He said he
couldn't wait to see me: "I've been thinking about you all day,
Marvelyn." I smiled in the dark. There's nothing better than
knowing that someone has been thinking about you when you're
not around.

I arrived and unveiled the pie. We both laughed, and then he
gave me a kiss and thanked me. As he ate it we sat on his couch
and just talked.

"How was your day, baby?" he asked, like always.

"Not too bad. What about you, baby?"

"It was all right, 'cause I knew you'd be coming home to me
eventually."

We talked like this for a while, flirting and laughing, the kind of talk that always leads up to something sexual. The pizza was shoved aside as we started kissing and getting closer. I was getting real turned on when he pulled back and said, "Marvelyn, I don't have a condom."

I was stunned. No dude in my entire life had just come out and been up front about it when he didn't have a condom. In my experience, most guys would just go until you called them on it or wait until the last possible moment to quietly admit they didn't have one and then passionately convince you why it didn't really matter anyway. Here he was, my Prince Charming, taking the time and care to let me know up front that he didn't have one.

Why would he be so clear? And then I realized—he must really love me. Why else would he announce it like that? "This is what grown-ups do, not like them high school boys. This is big," I thought to myself.

"I'm not going to pressure you or make you do anything you don't want to do," he went on. "If you want to have sex, you'll take off your clothes. If you don't, then you'll keep them on. That's how I'll know."

My mind was racing. "He knows I'm not on birth control. Does he see himself being with me forever? The worst that could happen is a baby. He must think I'm a good-enough woman to mother his child. This is coming from a guy with two jobs, a house, a car." It suddenly felt like a stamp of approval. And God knows approval is something I'd been looking for all of my damn life.

I sexily undressed, letting my clothes—and inhibitions—drop to the floor.

We continued to see each other pretty frequently. Sometimes I wouldn't see him for a week, but then we'd spend a few days together—talking and laughing and having sex. There was just one other time when he let me know—again, up front—that he didn't have a condom handy, and again I chose to make love to him anyway. I interpreted it as a symbol of how much he respected me.

What did I love about him? Well, there was the surface, of course: He was a man; he had his shit together. He treated me nice, cooked for me (well, if you classify microwaving as cooking), bought me the occasional gift, always said just the right thing at just the right time. He was a Leo, and I liked his take-charge attitude and his desire to be the center of attention. His cocky attitude made him even more attractive to me because he could back up all his talk. He really was doing his thing and doing it well. It made me proud to be with him.

But there was more to it than just the surface. I felt like he honored me. We connected—in conversation and in intimacy—on a deeper level than I'd ever connected with any other guy. He seemed to genuinely care about me and whatever family drama or money problems I faced. It wasn't just that he was listening until he could get into my pants. My Prince Charming was on my wavelength; he enjoyed being around me, getting to know me, sharing his fears and ideas with me.

I had my fingers crossed that, if I continued to show how cool I was—invested but not obsessed, present but not a pest—our relationship would go to the next level. I fantasized about having his baby, maybe moving in and putting my own Marvelyn signature style on the place, meeting his family, and maybe, even one day, getting married. I hoped we would get more serious soon.

I had no idea how serious things were about to get.

Miss Wanda's Daycare—
this is where I made my lip gloss money.

chapter seven

Life was so incredibly full that I started to get tired. Or at least that's how I perceived what was happening.

I was working two jobs—the day care in the early morning and the pizza place into the night. Since Cortney, Kendria, and Sha were out of school on summer break, the four of us could play together, enjoying Nashville nightlife. I was falling in love, trying to be with my Prince Charming as much as time (and he, to be honest) allowed, but also trying to keep everything casual and moving at a pace that made him comfortable.

Then, just like that, my body gave out. At first it was just fatigue. I'd try to get up in the morning when day care started, and I would feel glued to the sheets, as if I would have to pull myself from them with the force of my will if I wanted to get up and shower before heading downstairs to face the onslaught of kids.

There were times in high school when I hadn't wanted to get up, felt lazy and unmotivated to get to school or practice on time, but never had I felt this kind of exhaustion. It was palpable, like it had seeped into the marrow of my bones.

I chalked it up to being too busy, and tried to muscle my way through it. I would yell at myself in my head, "What's wrong with you? This is ridiculous!" Sometimes I even gave myself pep talks right out loud: "Come on, Marvelyn. Get movin'."

Then I started bargaining with myself—"All right, you don't

have to shower. Just get up, pull those sweats on, and get down-stairs."

That's when my aunt started to get worried. She knew that since my transformation from tomboy to pretty girl I had been paying a lot more attention to my clothes and hair, always making sure that I at least looked presentable. But now I was dragging myself down the stairs with my hair all a mess.

"You are not in Marvelyn mode, girl. What is going on?" asked my aunt lovingly, but I couldn't even muster a respectable answer. I just shrugged, as clueless as she was. Eventually she got worried and called my mom, telling her that she thought something might be wrong with me, that I seemed unusually tired all the time and wasn't taking as much pride in my appearance as I normally did. "Take her to get checked out," she urged my mom. "She's not going to go on her own. I even took her temperature, and it is a hundred and two. That's not normal!"

She was right. Despite being exhausted and unmotivated, I was still trying to get to both jobs, and even still trying to party. When I was out with Cortney and the girls, I was tired, but at least I knew I was really living. I hated the idea of sitting at home, watching TV and wasting my life away. I overrode my body's instincts, which were just to sleep, and forced myself to get to my usual spots. I didn't like going to the doctor and didn't figure I needed to. "It will pass," I kept telling myself.

My mom showed up and dragged me off to an emergency clinic run by Vanderbilt. She was convinced I was pregnant. I was dead set on denying it, though I was starting to suspect the same thing. Prince Charming and I had only had unprotected sex those two times, but as I knew from girlfriends and health class, two times were enough. I could easily be pregnant with his baby.

When the nurse said the test was negative, I was actually a little surprised (and one tiny tinge disappointed). "If I'm not pregnant," I thought to myself, "what the hell is wrong with me?"

It was almost as if the negative pregnancy test made my mom even more mad. She didn't have the satisfaction of saying "I told you so," of being disappointed and letting me know as much. We fought all the way back to my aunt's house. Between my icy silence and her unending screaming, nothing was solved. Nothing was accomplished. As usual, in my moment of need, my mom was too busy criticizing me to notice how much I needed her. And as usual, I was too busy shutting her out to know that I really needed her close.

When my mom dropped me off at my aunt's house, I slammed the car door behind me and went upstairs to the room where I slept, crashing into bed. Staring up at the ceiling, I thought about my situation.

I had always been healthy. Because I was an athlete, my body was my tool, so I had to keep it in good shape. I didn't always eat healthily, but I had never been overweight or prone to getting flus and colds. I prided myself on my tight, muscular physique, my strong arms, my tiny stomach. It was like my body had always been a machine—doing what I needed it to do, whether that was being the first across the finish line or the last to leave a club.

Now, all of the sudden, I couldn't trust it. It wasn't just that I was sleepy. I was *exhausted*. I also felt dizzy a lot of the time; at the day care, I'd pick up a child I'd picked up with ease before, and it would take a huge effort. I actually worried about falling down in front of one of the kids. I'd lost my appetite completely—the pizza at my job looked unappealing, even disgusting when I was feeling really tired and queasy. Pepsi, which I loved, didn't sound particularly good to me. I was losing weight fast, losing my nice ass and my curves.

The nurse at the clinic said that it must be some short-term virus or bug that was going around the city's youth and gave me some antibiotics, but in my heart I knew she was wrong. I knew there was something more serious going on.

I felt like I was disappearing, vanishing from sight. And there didn't seem to be anything I could do about it. No matter how much I tried to muscle through the fatigue, there it was again. I was beginning to understand that whatever this was, there was no escaping it.

It struck hard the next morning. After a real struggle, I had convinced my body to get out of bed and into the shower. I was feeling a little refreshed but still really weak as I headed down the stairs at about six o'clock to start the day's work.

Suddenly everything around me started to fade into one tiny circle of bright light, and then I was gone. Later I would learn that I had fainted on the stairs, tumbling down a few steps before landing—a pile of skin and bones where I had once been muscular and alert. My aunt found me and immediately started calling the family.

My grandmomma raced over and my uncle had to leave work to come take me to the hospital—this time Madison Minor Medical Center. The nurse took one look at me in the waiting room and immediately got me on IVs. It took hours to stabilize my body enough to rush me to the intensive care unit at Tennessee Christian Medical Center. My grandmother screamed, "Vanderbilt just sent her home yesterday! What is wrong with this child?" But the nurses couldn't answer any questions until some tests had been done.

Of course I don't remember any of this. I was in and out of consciousness but barely awake even when my eyes were open. Every once in a while I would see a new face that I had not seen in years. "This must be death," I thought. "I must be dying."

And I was ready. It felt as if God was giving me an out—a way to fade from the world without being responsible. A way to quietly, quickly slip away.

My mom started making the funeral arrangements—finding a place for the services, searching for a photo that would be suitable for the obituary. Relatives and old friends cycled in and out, some of them even saying good-bye as they left as if it was the last time they would ever see me alive. My uncle and mom missed work, my aunt would race over after the day care closed, my sister Tab only came alone so that the kids would not have to see me the way I was, everyone was making every effort to be at the hospital as much as possible. Even my daddy showed up and didn't leave my side for days on end. Everyone seemed convinced that I was dying.

But it would turn out that I wasn't fading as fast now that I was in the ICU. The doctors and nurses assigned to my case just kept scratching their heads, doing more tests, coming up with new potential diagnoses, asking my momma and aunt a thousand questions about my health history and current lifestyle. None of it pointed in the obvious direction of any disease. They knew I had pneumonia, but they couldn't figure out how a totally healthy, athletic nineteen-year-old with no history of illness whatsoever went from being in top shape to dying within a matter of weeks. I'd barely had a common cold before this hit.

The doctor decided that I needed a spinal tap—a test that no one, no matter how sick, can help but stay awake during. Especially once their doctor says, as mine did, "If you move, you could be paralyzed."

She pulled out a needle the size of one of my three-and-a-half-inch stiletto heels, and for the first time in days I was hyper-alert. When she asked, "Are you okay? Are you okay?" over and over again, I had half a mind to scream back, "Damn, lady, does it look like I'm okay? I've got a huge needle stuck in my spinal column."

But she was too nice to be mad at. She kept squeezing my hand and telling me about her kids. I was feeling the pain big-time, but asking her questions helped distract me from it: "That's right, honey," she said, "just keep talking."

The pain was like nothing I'd ever experienced. The only thing that comes close is when you are on the top of a roller-coaster and you know it's about to drop and your whole body goes numb in anticipation of what you are about to experience. It was like I had that numbness, but then the pain was ten times worse than what I could have anticipated, unlike on a roller-coaster where the drop always turns out to be not quite as bad as you imagined it. I still hate roller-coasters.

When she pulled the needle out I felt a tremendous relief, and then something almost like regret. I realized that as difficult as the pain had been, it had also felt like some kind of outlet. As someone who had always been a major athlete, I ran my body hard in order to release stress, feel alive, reconnect with myself, but since getting so busy working and partying, and now getting so sick, I'd felt like I was outside of myself for months. That pain sort of put me back in. It reminded me that this body was mine.

When the spinal tap produced no unusual results, the doctors resorted to the really expensive test: the magnetic resonance imaging (MRI). To me it seemed like one big tanning bed (though I'd never been in one, I had an active imagination). I had to take out all of my bobby pins—there went my weave—and my tongue ring. "Damn," I thought, "now I'm going to have to go get it repierced"— never mind that I was getting a procedure done usually reserved to detect lethal cancers and other super-serious diseases.

Once slid inside, I felt totally trapped. It was—once again— just me, myself, and I. There was no way out. There was nothing I could do but just lie there and think: "What the hell is this test for? All this shit is beginning to scare me. I've never known anyone who got all these tests. Why do they keep coming back negative? What are these doctors looking for? What's really wrong with me? I have some kind of freakish, unknown disease. They just keep doing test after test . . ."

I was starting to panic, but I remembered that the doctors had warned me that if I moved, I would have to do the test all over again. The last thing I wanted was more time in this tiny tube. What I wanted was some answers—one way or the other, I just wanted to know if I was going to die.

I was tested for all types of cancer. Negative. Then there was the college-related illness: meningitis. Negative. After a week and a half in the ICU, after dozens of tests, no one knew what was wrong with me, and I was still deathly ill. My mom continued to plan my funeral.

Two weeks after entering the hospital, I finally learned my fate.

I was asleep when the doctor knocked on the door. "Come in," I said quietly, unable to muster more than a barely audible sound.

The door slowly opened, and he came in and sat down next to my bed. "How are you feeling?"

"All right," I said, my standard non-answer after two weeks in a hospital with no answers.

"So you know you have pneumonia?"

I nodded. They'd told me this much. We just didn't know what had caused it.

"Well, the infectious-disease specialists have some test results for us that I think we should go over."

"Okay," I said slowly, wondering what the hell this doctor was trying to get at.

"It turns out that you're HIV-positive, Marvelyn."

It was one of those moments—the kind where one life ends and another begins, the kind where you find out someone has died or someone is pregnant, or that your younger sibling is getting married and you are still single. But it's funny—no matter how many times I experience these kinds of moments, they never seem real—at least not at the time.

When the doctor said those three letters—*H* and *I* and *V*—followed by that word—*positive*—my heart didn't lurch. My stomach didn't drop. My mind was immediately ready to plunge into action: *So what does that really mean? What do I need to do for that? What sorts of life changes does this require?* The real question, the question that everyone else would immediately think—*Am I going to die?*—didn't even cross my mind.

I don't know why. I was more aware of the fact that this doctor was annoying me, talking so slowly, acting like I was a breakable doll, than I was of the fact that he had just informed me that I had an incurable sexually transmitted disease.

"Marvelyn? Marvelyn, are you okay?"

"Yeah, I'm fine," I answered.

"Usually people don't take this kind of news so . . . well . . . so, in stride. Are you sure you aren't in shock?"

What did he expect me to do? Throw something? Throw up?

Scream obscenities at him? "I'm fine," I repeated. "So, what do I do?"

"Well," he said slowly, getting up from his chair, "we can talk about the next steps later. There's actually a doctor on staff who is very interested in your case, since it appears to have been living inside of your body for such a short time."

So now I was a specimen for research. Great.

"But we'll talk about all that later. Right now I think you should just focus on getting used to the idea, taking it in, processing what it means for your life."

"Well, what *does* it mean?" I asked, getting more annoyed by the moment. Was this guy going to give me any information? How many times did he play rock-paper-scissors before he admitted defeat to the other interns and reluctantly agreed to be the one to tell me? "Could you bring me some information or something?"

"Information? Yes, absolutely. I'll bring some information right away," he said, inching toward the door. "I'm going to go get it, but will you promise me that you won't hurt yourself in the meantime?"

Hurt myself? I could hardly move. Had this guy seen the thrush all over my mouth? Had he seen how hard it was for me to even get myself out of bed and to the bathroom? "I won't hurt myself," I assured him.

"Okay, good. I'll be right back," he said, disappearing out the door.

"Good," I thought. "At least I know what's wrong with me. Now I just have to figure out what the hell HIV really is."

My big sister Tab.

Photograph courtesy of Fred Cowan

chapter eight

The second that the doctor closed the door, I reached for my cell phone on the bedside table. My instinct in situations I was nervous about was to call up a friend, usually Cortney. "Hey, Cortney," I said into the phone.

"Hey, girl. You know I can't talk long 'cause I'm at work, but how are you?" she said nonchalantly. I could hear the buzz of her workplace in the background.

"Same old, same old," I told her, "but the doctors finally figured out what's wrong with me."

"That's great. What is it?"

"They said I have HIV."

I could hear Cortney breathe into the phone as she paused, an uncharacteristic silence on her part. "Marv, I can't talk right now. I'm at work. As soon as I get off I'm going to come straight to the hospital so we can talk," she said, her tone totally different.

"Okay, cool," I responded, then hung up the phone.

I sat in the hospital bed, cell in hand, and thought about her reaction. Not what I'd expected. Not what I'd hoped for. When I first heard the doctor say "HIV," it seemed worrisome but nothing too crazy. I was hoping I would call up some friends and hear reassuring responses that would set my mind at ease—"Ah, Marvelyn, you'll be fine, girl. It's okay. We'll work through it."

But there was no mistaking Cortney's tone. She was worried.

I know her better than she knows herself. So now I was worried. I decided to make more phone calls until I heard that reassuring voice I was looking for.

The next person who came to mind was Ashley. I'd known her since high school, but we'd only recently gotten close. So close, in fact, that she asked me to be her unborn baby's godmother when the time came. She was eight months pregnant, and we were already planning her baby shower. I was thrilled at the prospect of being her child's spiritual guide and already had items on layaway and had given Ashley lots of money.

"Hello," Ashley answered after just one ring.

"Hey, girl, I just found out some crazy news," I said, getting right to the point.

"What's up?"

"I'm HIV-positive."

After what seemed like ten full seconds of echoing silence, Ashley finally responded: "Okay, Marvelyn . . . I guess . . . I guess that I'll come see you." Her voice quivered in such a way that I knew, right then and there, that I would never be her child's godmother. I wanted every penny back.

It was starting to hit home for me that everything had changed. I had no idea what HIV was, but I was getting a sense that it was something that would distance me from people I cared about.

The more negative reactions I got, the more people I wanted to call. It was as if I truly believed that a nonchalant reaction would mean that HIV was a nonchalant disease. At that time in my life I was so hooked into others' perceptions of me, I counted them as gospel. I needed that stamp of approval.

I realize now that I was incredibly naïve back then. I didn't preface the news of my diagnosis with small talk. It never even occurred to me that this was something people needed to digest. I came right out and said, "I'm HIV-positive." Looking back, I realize I could have at least said something like "Do you have time to talk? Can we talk?" or considered the fact that they were on the job and told them later.

The next person that I just busted out with my news to was Tab, even though she was at work. I figured if friends were going to bug out, maybe my own flesh and blood would be able to take it in stride and help me understand what was going on.

Tab knew enough about HIV to panic. "My nerves are bad," she said. "Oh Jesus, I got to sit down. Bye, Marvelyn."

When the line went dead, I only hoped that my sister didn't burn anybody with the curlers. I decided I'd call my aunt. She always managed to avoid drama and approach situations with a level head. Plus, she'd always been in my corner. I figured this would be no different.

I figured right, but her reaction wasn't exactly levelheaded. "Marvelyn, this is really serious. Do you know what HIV is?"

"Naw, Auntie, that's what I'm telling you. They told me I had that shit and then they just bounced. They said they'd bring information so I can read about it, but they haven't gotten back yet."

"Sit tight. I'm going to come straight to the hospital when the kids leave the day care," she said. As I clicked the red button on my phone, I had to laugh to myself a bit. Sit tight? Where in the hell did she think I was going? I weighed barely 105 pounds and had an IV hanging from my hand. My newest accessory was a plastic ball with a cord right in the middle of my legs, better known as a catheter.

After the gravity of my sister's and my auntie's reactions, I realized I'd better call my momma. She'd be hurt if she found out through someone else. I dreaded her response.

Dread, it turned out, was an appropriate response. She inhaled deeply when I told her and then said, without skipping another beat, "Do not tell anyone anything, Marvelyn. If people want to know what is wrong with you, tell them you got cancer or something."

"Cancer?"

"Anything but HIV. I'll be there soon."

My first experience sharing my diagnosis had been a total disaster. Not only did I get none of the reassurance that I was hoping for, but I'd learned nothing more about what HIV was—other than that it must be pretty damn serious if my mom would prefer I tell people that I had cancer. Then I thought it was just another case of Marvelyn embarrassing the family, my mom's shame over my not living up to her expectations showing its ugly face again. Now I know that she was already anticipating the stigma attached to those living with HIV. She wanted to protect her baby. I had no idea, at the time, how much protection I would need.

I set the phone back on the bedside table and resolved not to call anyone else until I had a chance to learn more about HIV. I tried to reason out how I might have gotten it. "Must be those pre-schoolers," I thought, shaking my head. Their noses were always running. Their hands were always dirty. They coughed endlessly. Images of all the times I interacted with the kids flashed through my head—holding their hands, helping them clean up after snack time, changing their diapers. It seemed like the most likely time I could have gotten HIV.

The door slowly opened and the doctor peeked in tentatively, "Mind if I come in?"

"That's fine." They'd never asked me that before. Nurses and doctors just paraded in and out like they owned the place.

"I've brought you some brochures on HIV," the doctor said, laying what looked like a computer printout on my bedside stand. "It's not much, but it will get you started."

"Thanks," I said as I picked it up and started skimming it.

"There is a specialist here in the hospital who is really excited about . . . I mean interested in your case. She is the one who suggested the HIV test, and she's going to see you at her earliest convenience. She's one of the state's foremost experts on HIV and AIDS, and she's—"

"AIDS? Am I going to *die*?" I asked, my eyebrows raised.

"Well, HIV is the virus that may or may not lead to AIDS, so those who are expert in one are always expert in the other," he explained. "The brochure will help to clarify all of this." He was already inching toward the door, away from the minefield of explaining the connection between HIV and AIDS and, after that, the connection between AIDS and death, to the shrinking teenager before him.

The brochure was titled "HIV 101," and I quickly realized that it wasn't going to tell me anything about actually living with HIV. Instead it was a very clinical explanation about what HIV actually was, with a few statistics as well—also helpful, since I had no idea what those three letters stood for.

I read that the *H* in *HIV* stood for *human*. I felt dumb. Then I read that there was no cure for HIV, and I felt hopeless. As I continued reading, I learned that over 50 percent of people living with HIV die from stress-related causes such as depression and fatalism, and I thought for sure I was going to die. Thinking about it really made me stress out: "If this isn't stress, I don't know what is."

But how did I get it? The back of the page explained. HIV was transmitted primarily through the sharing of needles or unprotected sex with an infected person. I didn't do drugs and always made sure a brand-new clean needle was used when I got a tattoo. The only explanation was that I'd gotten it through one of my careless sexual encounters. There hadn't been many of them, but there were enough that I started shuffling through the faces in my head.

And then I looked at my phone. Clearly I had a responsibility to tell these guys so that they would go get tested. One of them had to have given it to me. As much drama as there had been in various relationships, I definitely didn't want any of the guys I'd slept with to go through what I was going through—being laid up in the hospital, at death's door with no answers. In fact, I still genuinely cared about most of them.

What's more, I knew that if I didn't tell them, someone else would. Nashville was a big city—statistically—but it usually felt like a small town when it came to how fast news traveled, good or bad. I didn't want Ashley's sister's cousin telling the guys I dated that I was going to die and they might too.

At just that moment, anger welled up inside of me. How did I not know that this virus was sexually transmitted? I felt I had been robbed, by my community, my school, and my church. The mantras I'd heard over and over again growing up—"Don't do drugs." "Don't get pregnant." "Don't smoke."—suddenly seemed so worthless. I had seen family members struggle through drug addiction and lung cancer, so that was all very real to me. I'd seen friends get pregnant at a young age and have their whole lives change, so that too was something I understood. But never had someone mentioned the possibility of me, Marvelyn Brown, contracting HIV from unprotected sex. I had seen it as something only Africans or gay men got.

The image of a shriveled, destitute African popped into my head; that had been the photo that accompanied the tiny blurb about AIDS I had seen on CNN. Even my teacher has barely referred to it. How misleading! It should have been a photo of someone who teenagers in Nashville, teenagers everywhere, could relate to so we'd know that we were at risk. I was that picture now.

At that moment I wanted to blame everybody—Planned Parenthood for giving me a Pap test that I basically thought was an STI screening, the illustrator of my health book for making me feel grateful that my great-great-great grandfather or grandmother got on the boat to America, and the Bush administration for pushing abstinence only, knowing damn well that x percent of teenagers were indeed having sex. I remember thinking, "Why isn't this plastered on walls somewhere? Everywhere? Why isn't it shouted from the rooftops, and on billboards, and on television? If someone had just told me. If someone had just taken the time to explain that there were incurable, deadly diseases that could invade my body if I had unprotected sex, I would have protected myself 100 percent of the time."

I wasn't going to rob the men I'd slept with of the opportunity to protect themselves, that is, if they weren't infected already. I picked up my cell again and started dialing. Thank goodness that I never knew how to let a man go because I had the numbers of all the guys I'd slept with—except my first, Derek—in my phone.

First up was the dude I tried to get pregnant with in order to keep him around, James.

"How you doing, Marvelyn? I heard you were up in the hospital."

"I'm HIV-positive," I told him, like pulling a Band-Aid off in one urgent grab.

"Damn, girl," he said, and then immediately started reminisc-

ing about our days together, as if the reality of HIV had been in-
stantaneously replaced with the fantasy of yesterday.

"Yeah, I remember," I said. "But did you hear me?"

"You act like somebody hears this on a daily basis. People don't
call me telling me they have HIV," he responded, getting defen-
sive. "I mean, I'm here for you for sure."

"I know, ex-baby, I know," I said. "Sorry to spring it on you. I
just wanted you to know."

With that I let the conversation slip back into remembering
and joking. Clearly he wasn't ready to talk about it, and I wasn't
going to force him to.

Next I called the guy that I paid to go to the prom with.

"Shit, girl, that's terrible. Maybe if I was man enough, this
would have never happened to you."

"What do you mean?" I asked, wondering if someone was
going to finally admit to me that he had HIV.

"Marv, you are a beautiful, sweet, and fun chick to be around,
but I was not ready for no commitment. Do you know who it is?"

"Naw," I said.

"Good luck finding him."

And the next.

"I'm HIV-positive."

"Damn, I'm sorry, Marv," he said. "That must be hard know-
ing that you going to die. I'm sorry to hear that. I'll keep you in
my prayers."

Not one of these guys, for one second, thought he had it. Only
one even realized it was sexually transmitted. How had all of my
generation missed this memo?

Just as I was wondering how the hell I got it if none of these
guys thought he had it, my family started showing up. In fact,

within the next hour, it seemed like my hospital room filled up from wall to wall with people who had heard the news and wanted to come see for themselves. Cortney and her family, my momma, my auntie, my sisters—everyone rushed over and started buzzing around me, as if there was anything that could be done.

The hospital phone started ringing off the hook with people—some close, some mere acquaintances—who had heard of "my death" and wanted to express their condolences. After explaining one too many times that her daughter, in fact, wasn't dead, my momma pulled the plug right out of the wall. Flowers started showing up at the hospital for me, none of which I could enjoy because I wasn't allowed to be around live plants. I sometimes wonder if that wasn't a blessing. I'm sure some of the cards expressed sorrow about my "passing." Two things about Nashville were starting to become crystal clear: Gossip travels fast, and no one had a clue what HIV really was.

I found out I was HIV-positive on July 17, 2003, and so did the rest of my town. I told five people and they told five people and then they told five people . . . and on and on until it seemed like the total population of Nashville knew. Cortney told me that word on the street was that the pretty girls had already gotten wind of my diagnosis and basically announced it at one of the local chicken-wing hang spots. I shuddered. Now there was no doubt that almost everyone who went to my high school would know, along with people who had never even met me. Thus began my identity as "that girl with HIV."

There was a knock on the door and a new doctor walked in. All my visitors instinctually wandered out into the lounge. They knew it was serious when a man or woman in a white coat walked in.

"Hi, Marvelyn, I'm Dr. Lisa Laya. Mind if I come in?" She spoke quickly and efficiently. She looked like a grown-up Dora the Explorer.

"I'm an HIV/AIDS specialist, and I tested you for HIV. I want you to know that we are able and ready to do whatever it takes to get you well again, and thereafter, get you medicated so you can live healthily with this disease."

"Live healthily?" I asked. "Isn't that impossible when you've got HIV?"

"Not really, Marvelyn. Advances in pharmaceutical research and therapies for HIV patients have really changed the face of the disease. People can live very long, fruitful lives with HIV these days. You're actually quite lucky."

"Lucky?" It seemed like everything that came out of this woman's mouth was exactly the opposite of what I was expecting.

"You're lucky because you were infected in this time, rather than ten or twenty years ago when the drugs were still largely experimental and often ineffective. And you're really lucky that we've diagnosed the disease so early in its development."

"More like blessed," I thought. "How early?" I asked, my mind racing.

"Marvelyn, HIV has only been living in your body for three weeks. I watched your HIV antibodies go from negative to positive over the course of twenty-four hours."

"Shit!" I screamed.

Now it was Dr. Laya's turn to ask questions. "What's the problem, Marvelyn? That is a good thing. You have a great chance of living a long and healthy life because we found out about your HIV status so early."

"I know. I am really happy about that. Honestly. It's just that knowing it's only been in my body that long means that I also know exactly who gave it to me," I said, shaking my head. I pulled out my calendar and looked at two dates in June, both with tiny hearts in the boxes that indicated I had sex with Prince Charming. I showed her the calendar and she confirmed what I had deduced. For the first time since being in the hospital, I was on the verge of tears.

"Ah, that can often be a real shock for people. Someone you really trusted?" The doctor asked, her face growing soft and understanding.

"Someone I more than trust. Someone I truly love," I said as I threw the calendar down and reached over and picked up my cell phone for what seemed like the millionth time that day. My fairy tale was fading fast as I realized that what Prince Charming had really brought into my life was a nightmare.

"I'll leave you to process this information, but I'll be back soon with more information about drug therapies," Dr. Laya said as she moved toward the door.

"Thanks, Doc," I said, managing a slight smile. From the time she walked in I knew that I was going to like this woman. When she told me she was the one who tested me for HIV, I knew she was going to be my primary doctor, and she still is. She seemed to have that rare mix in doctors—a lot of information and a lot of heart.

Prince Charming answered the phone on the second ring: "Hello."

"How you doing?" I asked out of habit.

"I'm good, baby girl. Did the doctors ever find out what was wrong with you?"

"Yeah."

"So, what's the problem?" he asked, hope in his voice.

"They told me that I'm HIV-positive," I said somberly.

There was a long pause, maybe ten full seconds of silence. And then he said what I knew he would: "I'm sorry, baby. I'm sorry. I got to go now."

There wasn't an ounce of surprise in his charming voice.

My bodyguard, my friend, my cousin Stan.

chapter nine

I was lying in my hospital bed, watching television, trying to zone out, when the phone rang. I didn't recognize the number on the caller ID, but what the hell?

"Hello," I said.

"How are you feeling, Marvelyn?"

"I'm cool. Who is this?"

"My name is Lynn. I'm HIV-positive, and I wanted to call and tell you my story."

At first I thought it was staged, that someone was playing games with me, but as her story unfolded, I realized she was for real. Later my mom confirmed it.

Lynn was a niece of my mom's longtime friend. My mom had asked Lynn to call to help me understand that other people with HIV and AIDS were living full lives. It was touching. At that point, I still thought my mom was just ashamed of me and my diagnosis, but she obviously wanted me to know that life wasn't over.

Lynn's story was inspiring. After six years of being in denial—just going to church and work and nowhere else—she started to tell her story. Now, eleven years after her diagnosis, she was a well-known activist, wife, and stepmother. I knew I was capable of what she had accomplished, but it was going to take some time.

Lynn was living in California, so I wouldn't get to meet her, but over the next year she would become my angel on the line.

The morning before I left the hospital, finally healed from the pneumonia and the painful thrush that had followed, Dr. Laya came to visit me for one last consult before my discharge.

"How are you feeling, Marvelyn?" she asked, sitting on the edge of my bed.

"I'm all right, Doc," I answered. "I definitely feel better than I did when I came in this place."

"Well, I should hope so," she said, chuckling. "You know I can refer you to a really excellent doctor for continuing care. You need someone good to manage your HIV symptoms and medications."

"Refer me? Girl, you're not referring me anywhere," I told her immediately. "You told me I had HIV, and you are stuck with me. I trust you!"

A smile spread across her face. I could tell that even though Dr. Laya was interested in my case because of its unusual nature—such early detection and diagnosis—she also really dug me as a person. Later she would tell me it was my sense of humor and seemingly indistinguishable spirit that attracted her to treating me, even though she didn't tend to treat individual patients on a long-term basis.

I felt a little like Dr. Laya's guinea pig, but I liked her anyway. She was one of the only doctors in the hospital who had given me the information straight and simple, no prettying it up or calling in social workers to explain the hard parts. Like me, Dr. Laya believed that tough truths were best communicated fast and unexpected, like a loose tooth pulled from an unsuspecting mouth. She didn't treat me like I was breakable, something I couldn't stand. Little did I know then that that's how most people would react.

"Marvelyn, you take care of yourself, all right?" she said,

stretching to standing and starting to move toward the door. "Sometimes it can be really tough to transition back into the real world after you have a diagnosis like HIV."

"It's all good, Doc," I assured her.

I had no idea.

My cousin Stan came to pick me up. Stan, a recon scout in the army, looked like the quintessential solider—strong arms, thick neck, cropped hair. He had always been a big supporter. When the hospital staff had demanded that Stan wear a face mask and gloves to visit me, he had gotten heated. "This is my baby cousin you're talking about! I'm not afraid of her. I'm not going to treat her like a leper!" I heard him shout. The staff had to explain to him that it was for *my* benefit, not to protect him. (I was especially susceptible to infections since my immune system was so impaired.) My heart was warmed by the idea that Stan wasn't afraid of me, regardless of my diagnosis, and that he would do anything to let me know that.

I was so grateful when he showed up on that sunny Nashville day to take me home. I had been in the hospital for two long weeks. My whole life—not to mention my body—had been turned upside down. My privacy had been invaded. I'd been poked, prodded, pricked, tested, and made to swallow pill after pill. I hadn't been able to dress in real clothes or get my hair done. I hadn't even been able to control who came in and out of my room. Sometimes it had felt like Grand Central station, or worse, like I was the latest in Nashville tourist attractions. Everyone wanted to come and see the girl who had HIV.

Except Prince Charming, that is. We hadn't talked since our brief phone call on the day of my diagnosis. My aunt found out how long the disease had been living in my body and jumped to the logical conclusion; she knew I was sleeping with Prince Charming and had been for months, so it only made sense. I wasn't going to confirm it, but I also wasn't going to deny it.

Before I knew it, my entire family was aware of who had given their little Marvelyn a deadly disease. The men—Stan in particular—were ready to march over to Prince Charming's house and kick his ass, but I begged them not to. My mom immediately blamed my auntie for playing a role in introducing me to him in the first place. The truth was that my auntie had little to do with it; she'd only invited me along to the park that day. But my mom would jump on any excuse that might make her feel less guilty about what was happening to me. I think deep down she somehow thought she was to blame.

I knew, even then, that blame wasn't going to get me anywhere. I begged my family not to do anything to Prince Charming. If anything, I thought he might need my help. Clearly he wasn't getting the medical attention he needed, and besides, he was the only other person I knew on earth who had HIV. The last thing I wanted to do was lose him. A tiny part of me even believed that this might be the thing that would push our relationship over the edge into marriage. If we both had HIV, and he had given it to me, wouldn't he want to spend the rest of his life with me so we could take care of each other and experience our shared fates side by side?

"You ready, Marvelyn?" asked Stan, interrupting my thoughts.

Going home would allow me to take back control over my life. At the very least, I couldn't wait to get my hair done and put on a decent outfit. *"Hell, yeah!"* I said, as he wheeled me out the sliding glass doors of Tennessee Christian Hospital. As the sunshine flooded over my shrunken body and practically blinded me, so did an undeniable truth: My life would never be the same.

As soon as I got home and Stan took off, I jumped into my car and drove straight over to Prince Charming's house. I knew my family would be livid, but there was nothing I wanted more than to lie in his arms and be told that everything would be okay. I thought that if anyone on earth would understand what I was going through, it would be him.

As soon as he opened the door and saw me standing there, he opened his arms and I fell into them. We went to the bedroom and lay down on his bed. I nestled my face in his muscular chest, taking in his familiar smell—like soap—feeling the softness of his T-shirt. It was like being enveloped in safety. Initially, we barely talked. I didn't have the energy to find out what he'd been up to, why he hadn't called, and he didn't ask any questions. We just listened to the air and enjoyed the sensual experience of being near each other again after all these hard weeks apart.

After finally opening up and talking about the elephant in the room, we were back to our old ways, kissing, getting involved in our usual foreplay. And right when we were about to have sex, he got up and went to the bathroom, coming back with a condom.

That condom in his hand felt like a knife in my heart. "After all this?" I thought. After he had infected me with HIV, after weeks of lying in the hospital with no diagnosis, wasting away, surrounded by pity and pain? After he had basically admitted to me that he had known he had HIV but had been in denial? After he had admitted that *my* diagnosis finally made his real? *Now* he was going to use a condom? *Now* he was going to play it safe and responsible?

All the anger that I had been too tired to feel earlier seemed to well up inside of me, but I still hoped that we could make our relationship work. I just wanted to feel loved. I just wanted to be held. I knew he was in denial. Against my better judgment, I went ahead.

Afterward, I put my clothes back on and left sadly. So much for an oasis.

At first I just moved back into my aunt's house and lay low. I was still pretty weak from the illness and the medication. The couch was about the only thing that looked good to me in that first week or two back in the real world. I watched *106 & Park,* talked on my cell to Cortney, and, to be honest, didn't do much else. I definitely didn't spend much time dwelling on my diagnosis.

Don't get me wrong. Just about every morning, the first thing that popped into my mind was, "I have HIV." It was a rude awakening over and over again—I'd force my eyelids open, see the sun creeping through the slit in the bedroom curtains, take a little stretch, and then it would hit me again, every time as hard as the first. I savored those first three seconds or so when I was just back to being Marvelyn, no death sentence, no shame.

I tried to live the rest of the day as if I was still that girl, but I wasn't fooling anyone, least of all myself. My family made it pretty much impossible. It was like everyone was afraid that I would break at the smallest touch. One little cough and my grandmother would call from the other room, "Do you need some soup, Marvelyn?" One little sneeze and my auntie would rush around the room like a banshee wiping every surface down.

I'd passed the HIV brochures on to them, but none of us really trusted the information we'd read; looking back, I think we were all too stunned to really process what it meant. As a result, we all spent those early weeks on tiptoes. I didn't want to make anyone feel unsafe, so I ate off of paper plates and washed my laundry separately. It made me feel like a leper, but that was better than feeling like a nuisance. After so many years of bouncing around to different relatives' houses, I'd gotten good at causing as little stress as possible for those around me. My new diagnosis demanded that I got even better at making other people feel comfortable with my presence.

The first time I returned to church was a real wake-up call. My arm hooked around Cortney's, I took a deep breath right outside the huge wooden doors, then let it out slowly as we entered. I imagined myself gliding in, just like before. I wanted to believe that God's house was one where I might feel safe again, where I might find the solace I hadn't found in Prince Charming's arms.

But the experience would turn out to be more about sensation and sin than safety. All heads turned as we entered, and then some just stayed there. I was an object of curiosity, just like in the hospital, but this time the eyes that stared at me were those of my enemies, not my friends or relatives. The pretty girls were posted up in the back row, looking on with smug pity as I made my way into a pew and quietly sat down. I knew that my diagnosis must make for great dish sessions and joke telling among them. It made my stomach churn.

I tried to just look forward and concentrate on the sermon. I hadn't been talking to God much—mostly because I couldn't even imagine what to ask of him but Why? Why? Why? I was too smart to pray for a miracle. I had HIV. Case closed. There was no changing that. So what else could I possibly pray for?

After that final betrayal by Prince Charming, I had become desperate with the idea of finding a man, settling down, and having kids. Rather than regretting the unprotected sex I'd had because it had given me HIV, I wished I had more of it. If only I'd gotten pregnant! Then I would already have a family started. My diagnosis made me feel like everything had to happen in fast forward. If I didn't find a man soon, if I didn't have a baby immediately, I would be dead. I would never have the experience of being a mother, and now that that didn't seem likely (What man would want a wife with HIV?), it was all I wanted in the world.

Of course I know now that those who are diagnosed with HIV in this day and age can live a long, healthy life, but at this point I was still surrounded by misperceptions and fear. I had read the pamphlets, but I hadn't asked any questions or done any further research to find out about the downside as well as the upside. I was afraid of what I'd find out. It seemed easier living in ignorance, with my own fears, than finding out the truth about what I could expect from my life.

I tried to pray: "Lord, I know that for so many years I hated myself and I wanted to do nothing more than die. I know that this is what I asked for in so many ways. But please, before you take me, Lord, just let me leave some legacy in the world, let some part of me carry over." That something, I thought, must be a child. I didn't have many role models who gave the world much else.

It felt good to be talking to God again, but I decided that I would steer clear of church until the town gossips had their fill of me and had moved on to some other sad story. I did need to go back to school. Without being officially enrolled I would lose my health insurance, and I really needed that now. I hadn't started my HIV meds yet, but I knew that they would cost a fortune. Dr. Laya had warned me to make sure I had my health insurance in order so that when she started me on the medication regimen I would be able to handle it financially.

There was no way, however, to prepare for how it would affect me physically and emotionally.

From an outsider's perspective, my fall semester may have resembled the one before it—classes I wasn't that interested in, teachers who droned on and on but didn't seem to care much about what they were communicating, homework I felt unmotivated to do. School had never been my forte, granted, but the first time I walked into a classroom after my diagnosis, I realized that I was encountering a new kind of hell.

Pssst . . . is that the girl with AIDS?

Hey, I think that's the one who's got HIV. Why is she in school trying to plan for the future if she ain't got one?

This teacher better not pair me up with her for any of the assignments. I don't want to catch some crazy disease.

They spoke in voices barely above a whisper, as if, along with my health, I'd lost my hearing too. But I heard them. I heard every word. And though I tried to face forward and keep my chin up, it broke my heart over and over again. The year before, I'd been the life of the party, a popular girl with lots of charisma, someone the others looked for when they wanted to know which club was popping off that weekend or where the house party was, or if they needed a shoulder to cry on.

A diagnosis later and I was an outcast in every sense of the word. In fact, the only girl who would be my lab partner was widely known to have herpes. We bonded over getting dirty looks from strangers in the hallways and sideways glances when we reached for silverware in the cafeteria, as if we were infecting everything around us with every touch. It was like living as a leper in the Middle Ages, right in my hometown in the year 2003.

I was psychologically shattered, but what made the transition back to school even more unbearable was that I was physically deteriorating too. Dr. Laya had started me on my medications—two giant white powdery pills twice every day—and warned me that there might be some side effects: nausea, dry skin, vomiting, diarrhea. There was no way, however, that she could have prepared me for how bad those side effects would be. In fact, I wondered why the word *effect* was even accompanied by the word *side*; there was nothing beside the point about them.

The bathroom became my best friend. I would go in there to take my pills, knowing full well that the second I swallowed them, they would come back out through one end of my body or the other (sometimes both at the same time). I tried to wait until there weren't any other students in there. I hated the idea of people listening to me gag; I feared it would confirm the idea that I was, indeed, a leper.

When my body started to adjust to the medication, I experimented with taking the pills outside of the bathroom. Sometimes I'd be okay—my stomach might grumble a little, but it wouldn't totally revolt. And sometimes I would immediately feel like I was going to explode. I would barely have time to get to a toilet, much less wipe off the seat, before my body would rebel violently against the medication that was supposed to be saving it.

One time, as I was leaving a class, a girl I had been friends with the previous semester leaned near me as I walked down the aisle of desks and whispered in my direction, "Serves you right, slut." I was usually pretty thick-skinned. After all, between my mother's criticism, the pretty girls, and all the other drama I'd experienced in my life, I had gotten used to handling insults. But this just felt like the straw that broke the camel's back.

I knew I wasn't a slut. "I just did what your momma did," I

thought. I'd had just as much sex as most of these girls, prob-
ably less than many of them, but they felt justified in pointing the
finger at me because I had the bad luck of sleeping with a villain
disguised as a prince.

It wasn't like people criticized girls around town who got
pregnant for sleeping around. They were just fulfilling everyone's
expectation of black girls from Nashville. No one thought twice
about what their irresponsibility—and damn right, I admitted I'd
been irresponsible—might mean for their future or that of their
kids, but my big mistake seemed to be fair game for everyone's
scorn. The shame stung me to the core, but deep down, I think I
was also so burned by my deepest knowing that the treatment I
was getting was hypocritical and unjust.

I ran crying into the cafeteria, thinking that since it wasn't
a mealtime I would be able to go into a corner and gather my
thoughts. Soon enough, however, I felt a hand on my back. "You
all right, sweetie?"

I turned around and was greeted by the big lopsided smile of
my psychology teacher, Sydney Hardaway. He was African Ameri-
can and always wore nice fitted suits. "Having a bad day?" he
asked.

"Yeah, I guess you could say that," I told him. "More like a
couple of bad months."

"Want to talk about it?" he asked, as sweet as the peach cobbler
that the cafeteria had served that day.

And just like that, my story came rushing out. It was the first
time I'd told anyone what had happened to me, what I'd been ex-
periencing since my diagnosis, the pain and the anguish, the un-
certainty and the confusion, the absolute, echoing loneliness. I'd

become very quiet ever since that day in the hospital room when my fate was sealed. My silence seemed like a weapon against all the rumors and speculation. But that day, in that huge, antiseptic-smelling cafeteria, my silence was broken by a flood of pain I finally acknowledged.

Professor Hardaway just nodded and rubbed my back repeatedly as tears welled up in my dark hazel eyes. In retrospect, I imagine he was quite scared of my diagnosis himself. No one in Nashville seemed to know about HIV and AIDS and how it is transmitted. But something allowed him, in that moment, to listen to this sobbing, skinny girl talk about the ways in which her life was crumbling and only reflect empathy and acceptance.

Angels come in mysterious forms. That's one thing I've learned from having this disease. That day, the divine came to me in the form of a middle-aged psychology teacher in a tight suit. He gave me the first real opportunity to tell my story, something I wouldn't do again for many months, but the seed had—without a doubt—been planted in that cafeteria that day. I will be forever grateful to him.

No matter how much I read the literature or talked to Dr. Laya, I was pretty much convinced that I was going to die shortly. It was just a feeling that welled up in me every time I thought about my diagnosis. I felt sick, like my insides were rotting, no matter how many horse pills I stuffed down my throat and tried to keep down. Now I know it was my hope rotting, but at the time I was convinced that my body had only a matter of months or a couple of years at the most.

At this point I was still staying with my auntie, but I was feeling pretty alienated from everyone. No one could possibly understand what it was like to have HIV. Cortney tried to be sympathetic, but she wasn't sure how to approach my problem. Prince Charming had offended me beyond reconciliation (or so I thought). My family tried to be helpful, but they only managed to make me feel smothered or patronized. I just wanted to be treated as an adult, to have some control over my life and my fate (which I assumed would be short).

So one morning after leaving school early due to heavy vomiting, without thinking it through beforehand, I pulled into the Super Wal-Mart. I parked my car near the front, walked through the automatic double doors as I had so many times before, and went straight toward the photography studio.

"I'd like to have my picture taken," I told the man behind the counter.

He raised his eyebrows at me in two little arcs and asked nosily, "What's the occasion, young lady?" I had dressed in one of my favorite American Eagle outfits—a cream-colored sweater and faded blue jeans that Tab had given me; I guess I didn't look like I was prepared for a graduation or engagement photo.

"I just want my picture taken, okay?" I said, not about to discuss my intentions with this perfect stranger. I'd had a hard enough time with the reactions of my own family and friends.

"Sure, darlin', whatever you say," he said, shaking his head as he led me behind the partition to the area where the photos were taken. He offered me a few different backgrounds—forest, ocean, and neon colors flashing like lightning—but I chose something plain that seemed simple and classy, representative of who I was. I posed lying on a mirror.

He took a few shots and then examined them on his computer monitor. "You sure are a photogenic young lady. I think this one will do just fine," he said, tapping the screen with the eraser of his pencil.

As I headed back out those double doors, proofs in my hand, I smiled for the first time in what felt like months. I may have been dying, but at least I now knew that my obituary picture would be beautiful.

My "obituary" picture.

chapter ten

Prince Charming called me on my cell a couple of months after my diagnosis. When I saw his number flash in the caller ID, I was tempted to just ignore it. I was still heated that he had insisted on wearing a condom after infecting me. Plus, he simply hadn't been there for me at all following my diagnosis. Any notion I had that we would be in this together, countering the rumors and proving everyone wrong *together*, had quickly been erased by his months of silence.

Despite my better judgment, I picked up. "What do you want?" I asked.

"I need to talk," he said. "Marvelyn, I know I haven't been good to you, but I really need to talk."

"This better be good," I said. "What's up?"

"I went to the Health Department and I took a test. It was positive. I'm HIV-positive. I'm finally ready to face it for real."

I felt my anger melt away. "We will work through this," I said.

When people hear my story, they're often shocked that I wasn't more angry at Prince Charming from the start. It's hard to explain, but I saw the two of us as linked by fate. He was the only person I knew with HIV in Nashville. He was also my first true love. Even close to a year after my diagnosis, there was a part of me—a naïve part, I admit—that believed we would end up together, living happily ever after in a home with two athletic kids with southern drawls. I always thought he would see some maturity in the way I handled the whole thing and fall in love. When he hit rock bottom, I reasoned, he would come back to me, accept his status publicly, and we'd saunter off together into the sunset.

I never told him about my secret dream, but I made it clear that I saw our fates as linked. "You're stuck with me forever," I told him after he finally admitted that he had infected me. "You will call me every day and check up on me." And he did. Not only did he feel responsible—finally—but he knew I was protecting him in public.

I never told people that he was the one who had infected me. First of all, I knew all too well the pain of having perfect strangers know something *that* personal about you. But beyond that, I was starting to read the literature on HIV and realized that it was un-ethical to ever tell anyone's HIV status, including the person who infected you. He told me because he trusted me. Even if he was the devil's best friend, I've never been someone who gets people back with a taste of their own medicine. I kill people with kindness.

We hung out occasionally. "How are your T-cells?" I would ask him (T-cells are thymus cells that help your body fight off viruses and diseases), but he would immediately shut down. It was like he'd opened the door just to slam it again. I ended up walking on eggshells when I talked to him on the phone, trying not to mention HIV.

But as was always the case with him, things slowly turned shady. I realized that though he had admitted his HIV status to me, he wasn't telling anyone else—not his huge family, not his friends, not even his own momma. He could have been my companion in dealing with the stigma, but instead he was making it even worse for me by staying hush-hush about his own diagnosis. As far as the rest of Nashville knew, I was still the only freak.

I also started to hear about him sleeping around with various women. Apparently they were also oblivious to those three little letters. Not only did I experience that sharp pang of jealousy when I heard the rumors, but I started to worry about these other women. I almost felt like I was betraying them by not telling them. I knew he wasn't disclosing his status, but I wondered if he was even using condoms.

One afternoon my cell rang and the caller ID indicated that it was an old friend from high school. I didn't pick up. I didn't need any more pity calls or, worse, ridicule. But when she called three more times, I figured I better pick up. Usually people don't blow up a cell for no reason.

"Your Prince Charming is messing with my cousin, Marvelyn," she said. I was just silent at first. What did these people want from me?

"I know you ain't never told me this is the dude that infected you, but he is messing with my cousin, Marvelyn. Is it him?"

"What's going on here . . . all this drama," I said, trying to avoid the subject. "Your cousin should ask him."

"He's denying it," she said. "And my cousin is pregnant with his baby."

That was it. I couldn't stand being responsible for the knowledge that he might be infecting other women and now a baby. I couldn't stand that he got away with being HIV-positive and not dealing with any of the stigma that I was dealing with. And truthfully, I couldn't stand that he wasn't taking care of himself. I didn't do it to be spiteful or disrespectful.

I got out the white pages and looked up his mom's number. I recruited my friend Shatoya to be on the phone for moral support and in case I lost my nerve. Prince Charming's mom picked up after a couple of rings.

"Hello."

"Hello, ma'am," I said, already starting to cry. "I really hate to bother you, but I have to get something out of my system. I feel it is best for you to know so you can help your son. He has HIV."

Silence. Then she replied, cold as ice, "Thanks for the information," and hung up. In her voice I heard the uncertainty. I wasn't sure if she believed me, and I needed her to.

Even more worried than before, I called my mom and told her the situation. I encouraged her to call Prince Charming's mom and make sure she really understood. In the back of my mind I thought they might even be able to provide support for each other. After all, as hard as my diagnosis had been on me, it had been equally hard on my mom. No one ever wants to see her kid, even one she's got problems with, suffer as I had.

When word got around to Prince Charming that my mom had called his mom and outed him (he never knew I called in the first place), he was livid and swore off talking to me for a while. In the process of trying to save him, I'd lost him . . . yet again.

A few months after my diagnosis I was starting to understand that the key lesson from all of this drama was, you can only trust yourself. I didn't even feel comfortable trusting Cortney. It was something I'd had a sense of all my life—moving from house to house, putting up with teasing from classmates and with criticism—but now that I was HIV-positive and either an object of pity or a leper in most people's eyes, there was no mistaking the truth. The only person Marvelyn could count on was Marvelyn.

School was becoming unbearable. I'd lost weight because every other time I took my medication, I threw it up. To make matters worse, my eyes were turning yellow because the pills were attacking my liver. My physical appearance was confirming all of the rumors.

Being around my family didn't feel much better. Everyone thought they knew what was best for me, to the point that I felt like a child. "Marvelyn, are you drinking enough water?" my auntie would ask.

Tab would double-check: "Marvelyn, did you take your medicine?"

"Marvelyn, let me make you some peas and carrots," my grandmother would insist if I picked up some fried chicken or drank some soda. It was enough to make me want to scream.

I had always been independent, always taken care of myself. To have everyone hovering around me and offering me advice they didn't really have the knowledge to give in the first place made me so heated. I was still eating off of paper plates at home, still washing my laundry separately. It seemed like it made them feel safe, and it made me feel better knowing they weren't put out by my presence. None of them had really made an effort to learn more about HIV and AIDS. I think they were just too scared of what they might find out. Ignorance is bliss, as they say.

But I didn't have the luxury of ignorance. I just had horse pills, some shattered dreams, and an urgent need to be on my own—away from the whispering college kids and hovering family members.

I couldn't afford to pay for an apartment. Hell, I didn't even have a job, so I did the only thing I could think of to get some privacy—I moved into my car. I would park at a twenty-four-hour Wal-Mart parking lot at night, recline the driver's seat, place my jacket over me, ball the keys in my hand just in case anyone tried to mess with me, and lean back.

I can still remember what it was like on those nights. I would look out the window, the huge parking-lot lights reflecting off the glass and illuminating the blacktop below, the sky wide and welcoming, and think about how the hell I had ended up this way.

It almost felt like my life had been cursed from the beginning, like I could never catch a break. First there had been my dad's disappearance, my mom's criticism, then being teased at school, the string of unfulfilling relationships, the fights with the pretty girls, then LaRena, my sweet, sad LaRena, then the failed ACTs . . . and then I'd met Prince Charming, and I'd really thought that my luck might be changing. In those short months of knowing him it was as if I was healing from all the heartache, starting to see myself as someone who deserved love and might even, if everything went right, *get* love.

My illness had been like a slap in the face, like the most haunting laughter mocking my naïve hope. Not only was Prince Charming not a sign that my luck was changing, he was the nail in my coffin. Sometimes, even with what I'd learned from LaRena's untimely death, I thought about just ending the pain myself. Why should I wait around for the AIDS to set in? Why should I wait around and watch my body deteriorate, my dreams die, everyone around me suffer? Dr. Laya said I could live a long life, but as my weight dwindled and my eyes turned yellow, I couldn't help but wonder if she was feeding me a line. I was not a fool. I saw evidence of the inevitable.

It was on those nights, staring out the window of my beat-up car in the Wal-Mart parking lot, that I came the closest to taking my own life. It didn't even seem like there was that much to take.

I woke up one morning, stretched my limbs as much as I could in the cramped car, and took a look in the rearview mirror. Only looking at my face and not into my eyes. This had been my ritual ever since I'd read about the facial lesions sometimes associated with HIV/AIDS. I was terrified of getting them. I was still struggling with medications. Dr. Laya had taken me off the two-pills-twice-a-day regimen, after seeing how sick it made me and how much weight I was losing, and now I took six pills twice a day. There were different side effects, which were no better. I began gaining weight in my stomach and some areas on my face. When I confirmed that my skin was clear, I took a big gulp of my warm Pepsi and started up the car.

Just as I was pulling out of the parking lot and heading toward the freeway, I remembered that I'd told my auntie I would call her. Nobody knew that I was sleeping in the car. They just thought that I was at a friend's house or wherever. I was still in touch with family members—mostly Tab, my auntie, my mom, and my grandmother. I would call them occasionally and let them know what I was up to—which wasn't much these days. I'd dropped out of school. If stress is one key factor in moving a person's status from HIV to AIDS, school wasn't for me.

I wanted to get a job, but I couldn't imagine what restaurant would want to hire someone with HIV or what clothing store wouldn't mind me touching its clothes. As it was, when I went shopping I never tried anything on anymore, knowing it would make the other shoppers uncomfortable.

"Hey, Auntie," I said, when she picked up after just a ring.

"How you doing, girl?" she asked, the concern evident in her voice.

"It's all good," I said, my official slogan now that I was trying to prevent anyone from worrying about me.

We kept chatting as I took the ramp onto the freeway. My

auntie gave me updates about the various kids I knew through her day care—loose teeth, sassing, and parents with a thousand excuses why they couldn't pay that month. She was too good to show any of the kids the door. I'd always admired the way she ran her business—smart, but always with the kids' best interests at heart. In a world with few role models my auntie was one of them.

I looked up from the radio that I had been fiddling with and suddenly realized that I was careening toward brake lights. I dropped the phone, grabbed the wheel, and tried to swerve out of the way, barely missing the car in front of me but going all the way into the next two lanes. I grabbed the wheel again, jerking it as hard I as could, and wound up turned around. A tractor trailer was heading toward me at sixty miles an hour. I watched his grill bear down on me, in shock, anticipating the end of my life in a way I never ever imagined.

The truck wheels started smoking—thick, black smoke, the smell of rubber burning—as the driver slammed on the brakes. As if in slow motion, those forty tons of metal and rubber came to a stop literally inches from my car. We were grill to grill, staring at each other with wide eyes through our respective windshields. The smoke continued to billow from his tires, creating the effect of a movie set.

I couldn't move. Even when the police arrived and knocked on my window, I was so shaken up that I had a hard time pressing the button to roll it down. "Ma'am, are you okay? Are you hurt?" the officer asked me, his adrenaline obviously pumping.

I looked down at my body, my emaciated body, and realized that I didn't have a scratch on me. "No, it's all good," I said, amazed that I could still utter those words. "It's all good."

I was blocking oncoming traffic. "We need you to turn your

car around and pull over to the side so that I can do a report," the
officer said.

I didn't want to get out of the car. I didn't want to move an
inch. It was like my whole body was clinging to this new home I'd
found, and this second chance I'd just received, a second chance
to live.

"I almost died. I almost died," I said inside of my head over
and over again. What if it wasn't HIV that was going to kill me?
What if life was so unpredictable that all of us, any of us, could
lose our lives any day? What if I wasn't especially cursed with bad
luck or doom but actually destined to do something important
with my time left on earth? What if God had just saved me for a
reason?

I hadn't thought about God fondly for months, but in that
moment I couldn't help but believe that he was speaking directly
through the slamming brakes and the billowing smoke of the
semi truck. He wanted something more from me. He didn't want
me to give up yet.

I had thought about death every day since I'd been diagnosed
with HIV. Every single day. And now I realized that it had been a
total waste of energy and time. Any of us—whether HIV-positive
or possessing a clean bill of health—is vulnerable. Any of us could
be living our last day on earth. I was shocked into realizing that I
wanted to make my life count for something.

In the weeks following my accident, I started really using God
again. I talked to him; I prayed that he would show me my pur-
pose. And slowly, piece by piece, I felt like I was putting together
the puzzle of what he wanted from my life.

When I was in the hospital, so many people said, "God wouldn't
give you this unless he thought you could handle it." At the time

I'd thought they were just making themselves feel better, comforting themselves in their guilt and pain. But after that nearly fatal car accident, I started believing them. There had to be a reason that I found out I was positive after this disease had only been living in my body for three weeks. It could not have been a fluke. I wouldn't have been given the opportunity to live longer, stronger—as a result of that early diagnosis—unless God had something planned for me.

And then, as if angels were dispatched directly after my epiphany, I learned about an organization called Nashville CARES, specifically designed to help locals with HIV and AIDS. Being in that office for the first time felt like a little slice of heaven. At Nashville CARES, everyone was highly educated about HIV/AIDS. No one was afraid to use the bathroom after me. No one avoided sitting on furniture I had just sat on. In fact, many of the workers—I would later find out—had HIV themselves.

My case manager, Marcia Williams, who is HIV-negative but cares about the cause deeply, welcomed me with open arms. She was a gifted listener. I came in scared and confused, with a lot on my mind. She did not interrupt. She did not tell me what to do. She just listened to all my problems about being homeless, scared to go to church and school, and she gained my trust right away.

She immediately helped me put things in place so that I could get a restaurant job and move out of my car. We decided it was best if I moved to a neighboring town, about thirty minutes outside of Nashville, so I could have a fresh start. I even chose to change my name—thinking it would help me remember that life was starting anew. From then on I was Shari—my middle name.

It's hard to explain how good it felt the first time I walked into my own apartment. I only had one door key and a trash bag full of clothes. I didn't have a single piece of furniture, not to mention toilet paper. I had nothing, and yet I felt rich. I lay on the floor and just cried and cried. I felt a sense of hope. Finally, I had a place of my own. Finally, I had some space to figure out how I was going to use my life for good.

Just days after I'd moved in, I received a call from another AIDS service organization in Nashville by the name of Street Works. They let me know that World AIDS Day was coming up and Middle Tennessee State University—my old stomping grounds—was looking for a speaker who would talk with the students about the realities of living with HIV/AIDS. "We think you'd be the perfect person," the executive director, Ron Crowder, explained. "You're right around these students' age. You're beautiful, funny, outspoken."

"Ah, stop," I said, feigning timidity. Before I could weigh the decision, I flashed back to my near car accident and that fateful day sitting in that ambulance, watching the smoke rise. "Sure, I'll do it," I said, then immediately regretted my decision.

How was I going to stand in front of all of those people, people I'd partied with, and admit to having HIV? How was I going to deal with all the humiliation that would no doubt follow? If the MTSU kids were anything like those back at my community college or high school, this was going to be an absolute nightmare.

But I'd said I'd do it. I wasn't about to disappoint any more people. And somehow I still had this feeling in the pit of my stomach that I also owed God.

One month later, I stood in front of a packed lecture hall at MTSU and stared out at the one hundred or so faces staring back. I had dressed in my old tomboy clothes, hoping it would prevent some of my old party crew from recognizing me (they'd known me as the Marvelyn in heels and tight clothes). It was a desperate move, but that morning when I'd been getting ready, desperation was about all I could feel.

I hadn't planned what to say. My case manager instructed me to just speak from the heart, but right then my heart was blank. Thirty seconds passed in absolute silence. I stared. They stared. Finally I just said: "My name is Marvelyn Brown. I am nineteen years old. I have HIV. Are there any questions?"

In fact, there were many. Two hours' worth.

How did you get it?

Did you use a condom?

What is HIV?

Do you still have sex?

Do you still date?

Didn't you used to go here?

It was a groundbreaking day, but to be honest, it freaked me out, because the audience was very receptive and attentive. All those questions confirmed what I had suspected ever since my accident—I was destined to speak out about HIV and AIDS. My mission was to tell people the truth about its transmission and wake them up to the fact that HIV was everywhere, especially in the black community.

While at Nashville CARES I'd learned that more than five thousand people lived in Nashville with HIV. Those numbers had shocked me, and I knew they needed to shock everyone else. I just wasn't quite sure yet that I was ready to be the messenger. I was finally aware of my power, and it scared the hell out of me.

My sister Tab's wedding with her new husband and their kids. **Photograph courtesy of Fred Cowan**

chapter eleven

You would have thought that first talk would have had me floating on cloud nine, ready to tell my story far and wide, but instead it sent me into hiding. All those questions, all those curious faces, all those desperate admissions afterward from students who were also HIV-positive (but not out about it, of course) had me reeling. I realized that my bravery could really change people's lives.

But first I needed to get my own life on track. Shortly after I'd moved into my new apartment, I was hit by a drunk driver and my car—a hoopty as it was—was officially totaled. Without a ride I was free from any official ties to my mom (she had still been paying my insurance), but stuck walking to work at Cracker Barrel, where I was waitressing and making good money.

That first week I made $550, or should I say Shari made $550, which I promptly spent on the cheapest car I could find over at the Tote the Note—a place that makes a killing off of people in a bind by loaning them money with huge interest rates without asking too many questions about their credit. That thing was pitiful, but it seemed to run.

Seemed being the operative word. I was good for about five miles until that damn thing broke down. And when I say broke down, I mean one of the wheels literally just up and fell off, rolling down the street like it had no sense. I called my mom, because *she* can screw with her kids, but *you* can't, and she threw such a fuss that the guy who'd sold me the lemon came and picked me up and put me in a fairly decent truck as a way of apologizing.

I loved riding around in that little truck, even if it was used. It made me feel like I was going places, figuring things out. I started filling my apartment with furniture; at first I couldn't afford much, so it was just Goodwill and garage-sale finds, with a few pots and pans from my auntie thrown in. Even my thirteen-inch television with no cable was something to be proud of, simply because it was mine and mine alone.

It was around this time that my sister Tab called to tell me some exciting news: "I'm getting married, Marvelyn!"

"Congratulations, girl!" I yelled into the phone. I already knew because her boyfriend, now fiancé, Emery, had told me.

"I got to ask you something though."

"What is it?"

"Will you be one of my bridesmaids?"

"Hell yeah I will!" I said, imagining myself in some terrible dress. I'd only really seen weddings in the movies, and the bridesmaids always seemed to be wearing god-awful dresses. I hoped Tab would have more taste. "When is it?"

"We're planning on July seventeenth," she said.

The picture of bad bridesmaid dresses in my head suddenly turned into an even more horrifying image—me, laid up in the hospital bed, wasting away. How could Tab not know that July 17 was the one-year anniversary of my diagnosis?

"Marvelyn, you hear me?" she asked.

"Yeah, yeah, I heard. Congratulations, big sis. I'm so happy for you."

* * *

Being solo was good for me—finally some peace and quiet after all those years of couch surfing and drama—but it also gave me a false sense of bravado. After months of six pills twice a day and upset stomachs, I decided to just stop taking my medication altogether. I started to wonder if I even had HIV after all. I wasn't losing weight anymore, and if I was, it was due to the medicine and not HIV. I didn't feel sick; I didn't look sick.

I thought about a few of the people to whom I'd confessed I was HIV-positive. Their responses were usually: *Shut up! Yeah right. Today is not April first.* I was right there with them. What if the doctor had made a mistake?

It was around that time that I started to subscribe to *POZ*, a publication specifically aimed at people with HIV and AIDS. I would lie on my Goodwill couch and read stories about people who had contracted HIV, the most recent medical therapies, discrimination cases, et cetera. In some ways, this helped me to be proud of myself for speaking out and consider doing more of it. But in another way, it actually fed some of my delusion. Most of the stories I saw in *POZ* were about gay men, leading me to wonder if maybe, just maybe, I had been misdiagnosed. I didn't look like these guys. I didn't have sex like these guys. What if the whole thing had been a huge, terrible mix-up?

The mix-up, of course, was all mine. A month after I stopped taking all of my medications, I went to see Dr. Laya. "You're not taking your medication, are you, Marvelyn?" she said, just moments after I arrived.

"How you figure?" I asked, shocked that she could tell when I didn't feel any different. I hadn't lost an ounce of weight as far as I could tell.

"Because you've lost four hundred T-cells," she said, her face dead serious.

My stomach dropped. It wasn't my weight I had to worry about losing. It was my T-cells. I would later learn that T-cells support a person's immune system in fighting off diseases and that untreated HIV kills these cells, leading to AIDS. A healthy, uninfected person usually has between eight hundred and twelve hundred T-cells. When an HIV-positive person's T-cell count drops below two hundred, she is vulnerable to developing AIDS. For these reasons, T-cells are like the positive person's holy grail.

"Do you realize that you are thirty-five T-cells away from getting full-blown AIDS?" Dr. Laya asked me.

I burst into tears. That was my last date with the delusion of invincibility. "What can I do?" I asked.

"I think we're going to try a new drug therapy. I understand how difficult it is for you to deal with all the side effects that go along with these things, but I think this new drug might be a bit easier to cope with," she explained. "I'm going to put you on Emtriva, Viread, and two protease inhibitors. That will equal six pills once a day."

"Okay, whatever you say, Doc," I said, still feeling shaky.

"But don't just say that, Marvelyn. It's actually best for you to *not* be on any drugs if you're not going to take them correctly. I need to know that you're really committed," she said, giving me a stern look.

"You have my word," I said. I was scared as hell, and nothing was going to keep me from taking my medication properly this time around. I was going to make it work. I was going to eat right. I was going to cut down on stress. Anything to get healthy.

In 2004, Gilead Science would combine Emtriva and Viread into one pill, called Truvada. That reduced me to five pills a day and only one co-pay a month. The new regimen would turn out to be far easier than I had anticipated because my body reacted so much better than it had to the other medications. I didn't vomit. I wasn't as tired. In fact, I barely had to run to the bathroom like the bad ol' days. I could waitress through a shift just like anyone else, go home to my increasingly decked-out apartment, and put my feet up. Truvada helped me live a normal life.

Cortney was working on her medical assistant degree at a local professional college and working at the Avon booth in the mall. Sha was a new mommy, and Kendria was still attending MTSU. I had friends at work, but we didn't hang out much outside of the restaurant just yet. In fact, sometimes it felt like the only people I really socialized with were my nieces and nephews. I loved having them over to make cakes and hang out at my place.

At this point I also managed to land a job at Olive Garden, where they served alcohol, which led to better tips. I was able to rent some of the finer things: a big screen, a nice bed, a whole set of pots and pans. I would get my *POZ* and *HIV Plus* magazines in the mail, all covered in brown paper as if they were pornography, tear off the wrapping, and spread them out on my coffee table, loud and proud.

Tab kept harping on me to shell out for my bridesmaid dress now that I had so much money, but I just couldn't bring myself to buy it. I couldn't process the fact that she'd planned her wedding on the only day of the year that had such totally devastating connotations for me. I tried to imagine myself standing up there, watching her walk down the aisle, but it was impossible to believe that I could do it with an open, happy heart. That day was just too powerful for me to forget.

Instead I spent my money on new clothes, little gifts for friends, and nice furniture. Looking back, I realize that I was sort of trying on pride at that time. My apartment and all its comforts were an attempt to create a little haven for myself where it was okay to be out about my status. I wanted to feel like even though I was HIV-positive and dealing with stigma from all the gossipers and drama queens back in Nashville, there was still hope for leading a somewhat normal life.

The trouble was that inside of my little paradise, I was solo. Though I was beginning to taste pride again, I was also sick of eating alone.

One day at work I waited on this decent-looking guy—young, a little shorter than I liked 'em, but charming—who obviously had a thing for me. We kept chatting when I came over to his table to check on things, and by the end of his meal he had asked me for my phone number. He was obviously impressed by my ride and, even more, by the fact that I had my own apartment at just nineteen years of age. "You're one of those independent women," he said, smiling at me. "I like that. You don't need no one to take care of you."

How true that was. We started chatting on the phone, but before things progressed too far I decided to test the waters as far as my HIV status was concerned. I didn't want anyone to think I was being misleading. One day as we talked on the phone and watched our respective TVs, he asked, "What are you doing?"

"Watching this AIDS special on TV," I answered, wondering what his reaction might be.

"Girl, there ain't no AIDS specials on TV right now. What channel you watching?" he asked, without skipping a beat. A good sign, I reasoned.

"That's so sad," I said, not answering.

"What?"

"This girl found out she was positive when she was just nineteen years old."

"What channel did you say you were on?" he asked.

"Actually, it's not on TV. It's on the phone," I said quietly, a little scared.

"Huh?"

"I'm positive. I found out when I was nineteen. I am nineteen," I said. I wanted to tell him that I was born with HIV or got it from a blood transfusion, like I had done with so many other guys before we dated. I hoped that they would feel sorry and try to accept me for something I had no control over. But I wasn't a victim anymore. I wanted no more secrets. No more lies. I just needed acceptance of me and what I had.

"Wow," he replied, obviously taking it all in. "We ain't got to talk about that again. I can tell you don't necessarily want to talk about it if you brought it up that way, so let's just put it behind us."

"Cool," I said, taking a huge, deep breath. I was so relieved.

Pretty soon we were basically living together. It was nice to have someone in the house, a living, breathing partner to share my new life with. He didn't have a steady job, so he would wake up in the morning and drive me to work (in my car). I liked the feeling of being dropped off and picked up, like I was being taken care of. I think he liked the feeling of taking care of someone, especially because he didn't really have the means to do it.

In truth, of course, it was me who was taking care of him. He wasn't making much money and I had plenty of it, so I would buy us groceries and pay all the bills and the rent. I hate to put it this way, but he really was my boy toy. Those high school days of me trying to buy a guy's affection were long gone. This time around it was more like I took care of him and got to benefit from his company and his caretaking. I didn't feel degraded. I felt in charge.

His momma even liked me. She thought it was great that her son had found such an independent, financially stable girl. After the drama with Prince Charming's mom, it was nice to feel accepted and even appreciated by my boyfriend's family. I knew, deep down, that he wasn't "the one" for me, but it just felt so damn good to have someone to wake up to, someone to come home to, someone to confirm that I was still beautiful and worthy of love every single day despite my HIV status.

All that, however, disappeared in the blink of an eye. One night he went out drinking with his friends and stumbled home sloppy drunk to his mom's house. The next morning he woke up throwing up like crazy, hungover as hell. It scared him, because he didn't remember drinking *that* much. He started to worry that he'd contracted HIV, not knowing that we hadn't even been together long enough for it to manifest, and what's more, we always used protection when we had sex. In a hungover, vomiting haze, he told his mom about his fears.

His mom immediately took him to the hospital, and then I got a very unpleasant phone call from her and her daughters: "If you infected my son with HIV, I will . . ." Well, you can imagine what followed—plenty of violent threats. I was madder than a mutt.

"If your son is positive," I screamed into the phone, "he didn't get it from me because we haven't even been together long enough for it to show up in his body, and second, he knew I was positive before he stepped one foot into my house. It wasn't no secret from day one!"

I never heard from his mother, or him, again. As much as it pissed me off, it also scared me. What guy wants to wear a condom for the rest of his life? Not only was I HIV-positive and stuck breaking that news to guys for the rest of my life, but now I realized that I essentially had to break the news to their families as well. Even if a guy accepted me, if his family didn't, I could lose him on the spot. And what mother really wanted to see her son walk down the aisle with an HIV-positive girl?

The worst part was that I knew he wasn't really my type. I wouldn't have dated him if it weren't for my status and, related to that, my loneliness. It had been nice to come home to him, to have somebody there, but I had really been lowering my standards. My self-esteem was shot.

One day Tab's fiancé called. "Marvelyn, I'm going to pay for your dress. The wedding is two months away, and Tab is freaking out," he told me.

"Thank you, I can't wait." But even as I said it, I knew I was lying. I couldn't stand thinking about that day, not only because it was the one-year anniversary of my diagnosis but because I was plain old jealous. I was starting to suspect that I'd never get married, never have a wedding like the one Tab was planning, with thirteen bridesmaids. I didn't have thirteen friends. It made me hopelessly sad. I told my grandmomma. I told my momma. Finally, I decided that I had to tell her. I knew she'd be mad, and that she had every right to be, but I had to break it to her.

I called her up, my hands shaking. "Hey, girl, what's up?" she answered, obviously unsuspecting of the bomb I was about to drop on her.

"Tab, I can't be in your wedding."

"What?"

"You want this day to be all about you, and as much as I would like it to be, no matter what you say, no matter what you do, it's not. Not for me."

She hung up, and we didn't talk again for eight months.

Around this time, a journalist from the local paper, *The Tennessean*, started blowing up my cell, trying to get an interview with me. At first I just ignored him. What did I want with more exposure? My reputation had already been dragged through the mud by all the haters in Nashville. They didn't need any more fodder for their rumors and lies. I wasn't going to give them the satisfaction of seeing my name in the same headline as those three letters.

But the more messages he left, the more I started to wonder. Could this finally be a way to tell *my* story *my* way, the *real* story, God's honest truth? I wasn't crazy about trusting some total stranger with it, but I hadn't had much luck with those who knew me. Maybe a journalist would get it right, talk about the huge infection rate among young women (I was learning, by this time, that though I felt alone, I most certainly was *not*; according to the Centers for Disease Control, black youth comprised the largest single group of young people affected by HIV).

I made an appointment or two with him, then chickened out when the time came. I felt so torn. Part of me thought it was a great opportunity to set the record straight, and the other part of me sensed that I would probably just be asking for trouble. While the two sides duked it out, this poor guy spent more than one lonely evening sitting at a table waiting for me to show up.

Rather than going to Tab's wedding, I spent my year anniversary at a family reunion in Indianapolis, Indiana. It was nice to be surrounded by aunties and uncles, cousins everywhere—all on my mom's side. I could be proud of myself with them. After all, I had a nice apartment, my own car, and a steady job. That was more than a lot of people my age could claim. So what if I also had HIV? My family wasn't about to bring that up.

They didn't have to. It was front and center in my mind and heart that day. Throughout the day I broke down, remembering those fateful weeks in the hospital and all that followed. I tried not to make a scene in front of family, so I'd find a quiet place and just get the tears out. My weepy reaction only confirmed the wisdom of my choice to not attend Tab's wedding. How could I be so sad on the happiest day of her life? I didn't want to ruin it for her, but I also didn't want to force anything; one thing about me is that I never take well to being forced into an emotion. Ultimately, I didn't regret not going. I knew it was something I just couldn't do.

It was a time of real reflection on the past and consideration of the future. One year after I'd been diagnosed with HIV, my life had changed in considerable ways. On the one hand, I was a lot more stable, seemingly headed in the right direction. I had all the trappings of a successful life—the job, the car, the home, the man (up until recently). For this I was truly proud of myself. If you had told me I would be so set up as I exited that hospital a year earlier, I never would have believed you. Then I'd thought I was destined for death at worst, loneliness and dysfunction at best. Instead I had a whole life of my own—independence, increasing health, even some hope.

But before I paint too pretty a picture, let me be clear. As much as I was proud of what I'd created, the success I'd built, the moves I'd made, I still felt empty inside. I wasn't speaking my truth in any significant way. After that first speech, I'd pretty much retreated inside of my little life, happy to be safe and private again. Being around my family for that reunion really made it hit home—people weren't any closer to understanding me or my disease. My own sister had planned her wedding on the anniversary of my diagnosis, signaling that she didn't understand what a significant experience it was for me. People asked about my job, my car, my apartment, but no one dared say, "So, how's your health, Marvelyn? Tell us about your HIV status."

I was still hidden, full of shame, and misunderstood. The newspaper article seemed like the first step in a long journey of changing that. When I asked my mom if she thought I should do it, she quickly said no. "Why bring even more attention to it, Marvelyn? Why not just lie low and do good like you've been doing? You're proving all the haters wrong with all your success." But my aim in life was not to prove the haters wrong. I didn't even want to stoop to their level.

I was just tired of not being able to express myself, not being able to sit down with my mom and let her know all the pain I was feeling, not being able to sit down with my sister and explain how hard it had been. In fact, I was pretty sure the whole wedding situation could have been resolved if I'd just felt comfortable talking, but I was all bottled up. I'd been keeping everything in, and I was tired of it.

Since I'd been diagnosed, people would sometimes ask me, "How you doing?" and I'd answer, "Dying." It was supposed to be a joke, an effort to play on their stereotypes of what being HIV-positive meant, but people didn't get that. They'd say, "Yeah, I heard." I knew they weren't purposefully ignorant, but they just didn't know. Slowly but surely, I was realizing that I wasn't the only one who hadn't been educated about HIV, and I damn sure wasn't the only one who had it.

I didn't want to disgrace my family or talk back to the gossipers. I just wanted to be heard. I just wanted to say something important. My goal was to live fiercely, say it like it is, keep it real. Deep down I knew that just lying low was never going to be my way. I was a fighter and a truth teller. I was ready to tell my story in my own words.

When I finally showed up for an appointment with the reporter at Logan's Roadhouse, I swear, he almost fell out of his chair. He obviously wasn't expecting me to really pull through. "Thanks for finally meeting me, Marvelyn," he said.

I liked the looks of him. He was bald, chocolate, and muscular. He reminded me of Alton from *The Real World: Las Vegas*. When we started talking, he didn't pull out a tape recorder or drill me with twenty questions. He just listened and took notes. He looked in my eyes and genuinely heard me. He asked some questions, but I asked questions too. After a nice meal and one of the most revealing conversations I'd had in over a year, I finally asked, "So hey, what's in this for you?"

He smiled mischievously and said, "It's the story of a lifetime."

The night before the story was finally going to run, I totally lost my mind. I broke down at my job. The general manager of the restaurant was aware that I was HIV-positive, because one day I became sick on the job, throwing up everywhere, and I had to tell her; besides, I didn't feel comfortable working there without her knowing that information.

I went into her office, closed the door, and we sat down and talked. I told her the article was going to run the next day. She assured me that I still had a job, but I still didn't feel secure. After all, in reality, she only paid me $2.13 an hour. Hell, she didn't have to pay me; the costumers did. I was nervous and immediately started training for the hostess position, where, if it came down to that, I wouldn't have to rely on the kindness of strangers.

That night when I got home, I called the reporter and threatened to sue him if he let the story go to press. Frantic and basically out of my mind, I yelled into his voice mail, "Don't do it! You'll ruin my life! You'll ruin my family! Don't do this to me."

But the story had already gone to the printer, and, even if it hadn't, I don't know that he would have stopped it. He knew, better than I did, that this could be a real turning point for me. He knew that I was going to help thousands of people with my truth, and he wasn't about to give up the opportunity to be the guy who made it happen.

I knew it wasn't his fault. I knew I was just fighting with myself. Part of me was brave as hell, the bigger part, I think, but there was a tiny part that wanted to crawl back into my cave and stay there forever, watching my big-screen television, serving as Shari, and letting it all go by.

Come six A.M., that possibility vanished.

"Shari, can you get table 210 ten for me? I am swamped! I got his drink order already, and he wants coffee," said my coworker. "Shari." No response. "Shari!" she repeated. I suddenly realized that she was talking to me.

"Yeah I got it," I replied. As I walked up to the table, I noticed that the man sitting there was reading the Sunday paper and turning to the Life section. The steaming hot coffee began to spill on my hand and all over my uniform.

"That is why you are supposed to carry a tray," said the manager. I composed myself as I walked toward the table. There it was—my face on the cover of *The Tennessean* Life section. The man couldn't possibly know who I was. I had changed my name to be unrecognizable, but I was still nervous that my identity would be revealed.

I put on my straight face and started to take his order. "Hello, sir. My name is Shari, and I will be taking care of you." I held it together enough to ask someone to take salad and breadsticks out to his table as I put his order in. I grabbed a copy of the newspaper from the bar and ran to the employee bathroom as fast as I could.

As I looked at the paper, I tried to fill myself with happy thoughts. "I'll be helping so many people and changing so many lives." But negative ones followed it. "Maybe that's just what the reporter told me so he could get the story!" A few tears snuck out of the corners of my eyes, but I immediately wiped them away as I looked at the first word in the title of the story: COURAGE.

I repeated it in my head over and over again. The reporter had worked on that story for months, and I figured that he knew me better than anyone else at that point. It took *courage* to do what I had just done, I reminded myself. I walked out of the bathroom to hear the hostess asking, "Who is Marvelyn? Different people have been calling all morning asking for a Marvelyn."

"Oh, that's me," I said, "Shari is just my middle name." Every time I had heard "Marvelyn" for the past year, I'd heard something that filled me with disgust and shame. I had started my new life an hour outside of Nashville as Shari. Shari didn't have to deal with the stigma of being HIV-positive. Shari got to glide through life, doing her thing, not worrying too much. But she also, I realized at just that moment, didn't have the courage that Marvelyn did.

I said, "I'm actually Marvelyn," then grabbed a wine bottle and proceeded to the next table with a huge smile of relief on my face.

Lynn's obituary; she was definitely my angel.

chapter twelve

Looking back I realize that having my story featured in *The Tenessean* was a turning point in my life. There is something about a public document that makes things undeniably real. Unlike my talk at MTSU, which was transformative but fleeting, the words in that newspaper were permanent. They signaled to the world that I was "coming out" as Marvelyn—a new Marvelyn who was not ashamed of herself or her status and, even more, intended to reclaim her life and make it mean something, not in spite of but *because* of those three little letters. I was a Marvelyn aching for something like self-love—not yet there, but moving towards it one brave and vulnerable step at a time.

Patrick Luther, the prevention and education supervisor over at Nashville CARES, contacted me after the piece came out and asked me to come in and speak with him. I'd seen Patrick around the office when I went in to get my own services, but we'd never exchanged much more than a glance and a quick smile. I didn't know what to expect, though I suspected it might have something to do with the newspaper article. After all, Patrick was used to dealing with the media and might have pointers for me about the kinds of things I should disclose and those I should keep to myself.

"I'm a little nervous," I told Patrick as I slid into the chair in his office.

"Marvelyn, let me paint a bigger picture for you," he said, skipping the small talk. "You have the potential to make a real difference in people's lives. You're well spoken. You're relatable. You're someone who people are going to listen to and pay attention to."

"Naw," I said, a little insecure.

"No, really, Marvelyn. You have to start taking yourself seriously. I'd like you to come work for us."

"Work for you?" I said, dumbfounded. I'd never thought about dedicating my life to working in the HIV and AIDS community. I figured I would put the story in *The Tennessean*, inspire some people to get tested, educate a few of the ignorant, and maybe, at the most, get a little pat on the back. But never did I think that I would create a career.

"For starters, let's say twenty hours a week working with one of our partners: the Comprehensive Care Center. It's a clinic for those who are positive."

I nodded like a bobble-head doll.

He continued, "That way you can keep up with your waitressing until you learn more. We'll train you as an HIV education specialist. You'll learn how to actually administer HIV tests, counsel those who are infected, point them in the direction of key services. We'll build your skills, and I, personally, will mentor you. I see so much potential in you for real leadership, Marvelyn."

Potential? Leadership? The words echoed in my head. Here was a man who had dedicated his life to eradicating misconceptions around HIV to prevent new infections and who made life more comfortable and healthy for those who already had it, and he was offering me mentorship. I immediately said, "I accept."

Looking back over my life, I realize that I was always scared to take charge, never feeling good or capable enough to inspire others. I quit the basketball team in high school because I was scared to lead my team to the play-offs. I was more comfortable being a follower and having the blame on someone else. Ultimately, I think my fear of leadership stemmed from my lack of self-love. I couldn't imagine other people accepting me, much less respecting me, if I couldn't even accept or respect myself.

But things changed after I got HIV. I had experienced the kind of isolation and hardship that makes the leader buried in a person step up real fast. Now that I had "come out" publicly, I had no choice but to be responsible, to take leadership over my own life. I had no choice but to learn to accept and love myself. Let's face the facts: It was that or death.

One of the hardest parts of my job was administering the Ora-Sure HIV tests. I can still taste the fear in my mouth as I would go through the motions—greeting the person, explaining the test, and assuring them that they'd have lots of services and support available to them should the results be positive. I would then direct them to take the stick and rub it on their gums, holding it there for five minutes, then let them choose from a selection of Jolly Rancher suckers because the testing stick was salty.

They had to wait three to five days, depending on the U.S. Postal System, for the results. Thank God, I never actually had to tell people they were positive, though so many of my colleagues were faced with that responsibility.

To be honest, I'd often shy away from testing people and let my coworkers do it instead. I couldn't imagine being the one to break the news, possibly because my experience had been so inadequate. I always jumped in if I could help explain things to someone after they'd been newly diagnosed. The initial moment scared the pants off of me, but everything after felt like territory I knew all too well.

At first I liked the work a lot. It gave me a good feeling—talking to people who were just like me, having coworkers who wanted to talk about more than what was on BET or MTV last night, having some real responsibility. I was still working at Olive Garden and making a lot of money, but I was starting to realize that I couldn't physically and mentally handle working both jobs.

By December I was moving into my very own office (with a door and a key!) for a full-time position at Nashville CARES. I would be doing a number of things with my promotion, including coordinating the First Person program, which got HIV-positive public speakers gigs throughout the Nashville area. I would also head up the Survivor Club, a program for high school students to teach their peers about HIV and AIDS.

It was a real step up in lots of ways. First of all, I finally had a solid salary. Waiting tables, you have no guarantees—you only hoped to make as much money as you did the week before. I also had my own health insurance. No more dealing with my mom's threats to take me off of hers or worrying that she would lose her health insurance and/or job because her daughter had an expensive disease.

Though it felt damn good to be independent, I was not quite ready to say good-bye to the cooks and busboys at Olive Garden. I had made a family with my coworkers. Nashville CARES was a place where I felt like I was contributing to changing the world, and it was a great feeling to be around other HIV-positive people, but Olive Garden helped me get my mind off of HIV and be around people my age.

That was part of the paradox of being HIV-positive, empowered, and yet still just twenty years old. I wanted to start what I knew would be an important career in HIV/AIDS education and prevention, but there were times when I just hungered to be a kid, to joke around with people my age, to forget—just for a moment—that those three letters defined so much of the rest of my life.

I think that so many HIV-positive people—especially those who work in HIV/AIDS education and prevention—struggle with this duality. On the one hand, we are so inspired to make a difference, to have our lives matter, to make those with HIV feel less alone and lead healthier lives and those who aren't infected to stay safe. On the other hand, we crave an existence that isn't totally defined by our disease.

The key, I've realized over the years, is to feed the second craving without being in denial. There's a way to hold HIV in your mind without letting it dominate your whole being. Sometimes you just need a break, which is fine, but you have to pay attention to the short distance between that and neglect of your health and harsh reality.

Having such an official position and so much responsibility wasn't easy. I was a damn good (and sexy, I might add) waitress, but this was a totally different kind of work. I didn't have any experience coordinating people, planning meetings and trainings, or keeping track of it all in spreadsheets and agendas—that type of stuff is boring to a twenty-year-old. And there was protocol—ways of doing things that were grounded in years of the organization's existence. I couldn't just freestyle everything as I'd been known to do in other areas of my life.

I've got to give it to myself, though, I learned pretty fast. I applied myself and asked for help when I needed it. I observed and innovated. And most of all, I did what I've always done best—I connected with people. I think the guests (we didn't like to call them clients because it felt too unfriendly) I worked with felt really comfortable in my presence, and my coworkers knew that. Though I may not have been the most experienced, I always worked hard and was usually good for a laugh along the way.

There was just so much to think about, so many people to keep in mind. Sometimes I was filled with self-doubt. Sometimes I wondered if I could really make it all happen. Fortunately, I felt Patrick believed in me (more than I believed in myself), and I would not dare let him down. His view of me—the leader with lots of potential, the well-spoken, caring young woman—helped me develop a more positive view of myself. Perhaps self-love is a seed often planted by another, though you have to grow it yourself.

A few weeks into working full-time, I received a phone call out of the blue from the National Association of People with AIDS (NAPWA), the oldest AIDS organization in the United States (based out of Maryland). Its mission, I learned, was to support those with AIDS and stop the pandemic. NAPWA also started the Ryan White National Youth Conference, named after a famous young man who contracted HIV through a blood transfusion to treat his hemophilia and went on to become a world-renowned activist and one of the first public faces of HIV/AIDS. Every year NAPWA would hold an awards luncheon to recognize young people making a difference in the world.

When I learned about what they were doing, I immediately told them, "I would love to join the host committee and help plan the Twelfth Annual Ryan White Youth Conference in my home-town." With people ages thirteen to twenty-four making up half of new HIV infections, I had to be a part of this movement. Besides, the idea of being around other young people with HIV was so appealing to me; I could hardly imagine how comforting it would be to look into the eyes of other positive youth, be included, and commiserate about it all.

One day while on a call discussing the conference, Aryka Chapman, an employee of NAPWA and the conference leader, said, "This year, for the first time ever, we will be giving out the Tarsha Durant Youth Award for Positive Leadership during our annual awards luncheon next month. Marvelyn, you are the recipient."

I was stunned—first Patrick giving me an opportunity with the job, now this! I could hardly wrap my mind around the idea of not only being supported but being celebrated. "I don't know what to say," I finally managed.

"A thank-you will suffice," she said.

I had gone from throwing up in the college bathrooms and daily humiliation, eating off of paper plates and washing my laundry separately, living in my car in a Wal-Mart parking lot, to really making a life and now a name for myself. Since I was a child I had always looked in the mirror and said, "My name is Marvelyn Brown and I am somebody." But now I could finally look in the mirror and say, "My name is Marvelyn Brown and I am somebody *with HIV*." Talk about self-love.

"Momma, I won some award," I told her over the phone the next day.

"What did you say, Marvelyn? You're cutting out." My mom had been traveling for months for work, so the last time I had seen her was to show her my new office, and we had only been in sporadic phone contact.

"I said I won an award," I repeated.

"What for?"

"It's called the Positive Youth Leadership Award. It's from this AIDS organization."

"Well, that's nice. What did you do to deserve that?"

"For telling my story. It's national, Momma," I said, still hoping she might actually congratulate me.

"Well, that's nice, Marvelyn. Congratulations." But it was hollow. Sometimes at moments like these you realize that it was never the word you craved but the feeling behind it.

"I would like you to come to the awards ceremony."

"I think I'm still out of town then. It's not a big deal, is it?"

"No, it's not a big deal," I said, then came up with some excuse to get off the phone.

I also kept calling Lynn, trying to tell her all the good news—about my job, my reward, my new self-acceptance, and the process of beginning to truly love myself for the first time in my life—but I couldn't seem to reach her. She was really the one who had first assured me that—once the grief over my diagnosis, the humiliation, the processing, the growing pains, were over—there would be a light at the end of the tunnel, that my life could mean something great, not in spite of HIV but because of it.

I was beginning to see a glimmer of that light and I wanted to share it with her. Little did I know that she had slipped into darkness.

When I walked into that ballroom at the Renaissance Hotel in downtown Nashville, I felt like an absolute princess, minus the puffy ball gown. I was wearing a pink and gray pinstriped suit, majority pink, and some gray square-toed pumps. Don't think power broker on Wall Street; think *Sex and the City* (one half Miranda, one half Carrie). It was youthful and hot.

The room was filled not only with flowers and decorations but with five hundred HIV activists, case managers, health care providers, and positive youth from all across the United States. Cortney and her family were beside me as we entered the banquet and found out we had a whole table reserved just for us. To my shock and enjoyment, my mother and her friends were already seated there.

I had underestimated my mom. I had thought she wouldn't understand the significance of the day, but seeing her there helped me understand that, though she wasn't always capable of expressing it, she had a lot of pride in the way my life was shaping up. Her initial disappointment over my diagnosis, and the fact that it would inevitably make my life hard, would never go away. But it could be complimented by a sense that her daughter was a survivor, the kind of young woman who made lemonade out of lemons—just as she had always strived to do despite life's struggles.

It was eerie—strangers were coming up and introducing themselves, congratulating me on all of my accomplishments (I didn't bother telling them that I didn't really feel like I had that many yet). I felt showered in praise. Even Ryan White's mother, Jeanne White, came up and expressed her admiration for my work and words. That was especially honoring, as I looked up to her for always supporting her son.

I thought that I was ready, but when the actual awards ceremony started, I was suddenly seized by the desire to be sitting on my back patio drinking a nice cold Pepsi—far from the limelight. I was nervous at the thought of standing in front of five hundred unfamiliar faces. I knew that I was among friends and a supportive community, but I didn't know what to say, what they might expect from me.

The announcer talked a bit about the article and how I was currently working at Nashville CARES. When she explained why I was chosen—my courage—it brought me to tears. "Please, come up and say a few words, Marvelyn," she finished, motioning for me to take the stage.

Something in me just took over and rose to the occasion, my nervousness melting away as I approached the stage and heard the roar of applause. I treated it like nothing less than the Oscars: "Thanks so much for this award. The last year has been a real struggle for me. I've been humiliated, abandoned, misunderstood, and made to feel very alone. But, right now, I feel the opposite of all of that. I feel surrounded by community and so, so blessed. Thanks to my best friend, Cortney. Thanks to my cousin who is in Iraq, and my nieces and nephew for keeping me young. Thanks, also, to my coworkers at Nashville CARES." There was not a dry eye in the room.

"Finally, I would also like to thank my mom and dad for giving me these good looks, and a special thanks to Hecht's for putting this suit on sale, because I look bad." Now everyone was crying *and* laughing.

The crowd went wild, standing up and cheering for Marvelyn, the girl with HIV. It felt like the fans cheering at the basketball game after I just hit the winning shot. I just stood there and smiled for a few seconds, taking it all in, looking at the award with my name on it, and then again at the smiling faces before me. For the first time since being diagnosed I felt entirely loved and accepted. "Thank you for this, God," I thought as I headed back to my seat.

After the luncheon was over, people kept coming up and talking to me:

Excuse me, Marvelyn. Can I talk to you?

Marvelyn, can I take a picture with you?

May I have your contact information?

You are so brave.

I was so flattered by all the feedback and inquiries. I gathered business cards while Cortney played manager, passing out mine. Finally, I posed for pictures with my friends and my family, holding the award proudly in front of my chest. I never wanted to forget this day or this feeling.

The conference that came after the awards luncheon was amazing. Children and teens from all over the nation got together to talk about their experiences of being affected and infected with HIV. It was also an opportunity to learn more about staying healthy and happy in a world that still misunderstands this disease.

I remember that first day, standing outside a huge conference room that NAPWA had reserved for the Positive Youth Institute, which was only for HIV-positive youth. I stood outside the door, nervous about entering. Nashville CARES had been a great way to meet others with HIV and gain a sense of community, but nothing could replace hanging out with HIV-positive people my own age. It was so good it almost seemed scary.

After finally entering, I looked around the room in awe. Youth ranging from eleven years old to twenty-five hung out everywhere. There were so many inspiring stories to learn.

I stood next to a beautiful girl who looked to be about twenty. As my gaze dropped from her face to her belly, I was shocked. She looked pregnant! In a thick New York accent she spit out at me, "What are you lookin' at?"

That was my first introduction to Maria, who would become

a good friend. She was infected with HIV at birth, but she didn't find out she was positive until she went to get tested for sexually transmitted diseases after being raped when she was nineteen. Her mother died at an early age and had not told anyone she had AIDS.

At the time of the conference, Maria was six months pregnant with her second child. I immediately wanted to know the how, what, when, where, and why. She was very open and honest, explaining to me that lots of HIV-positive women are successful in delivering negative babies thanks to meds, prenatal care, and cesarean sections. Even though I had heard and seen statistics confirming this, Maria was living proof.

Then I met Phillip, a sixteen-year-old guy from Washington, D.C. He had only known that he was positive for one month. He had snuck to the conference, telling his mom that he was over at a friend's house. He wasn't ready to disclose his status to her, fearing that she might kick him out or disown him.

I would soon find out that Phillip was full of secrets. He represented himself as heterosexual, though it was unprotected sex with a man that had infected him. However, I didn't throw the "down low" label on him. There is so much talk about how men, particularly African American men, are afraid to admit their homosexuality or bisexuality and, therefore, increase the HIV infection rate by having unsafe intercourse with both their secret lovers and their wives/girlfriends. It's an important conversation, but I hate that word. A person's sexual preference doesn't make a lick of difference (though I wish some of the men who do have homosexual or bisexual feelings could find their own self-love and stop pretending to be someone they're not). What matters is whether they have safe sex or not. Whether some guy is your boyfriend, your husband, or just a straight-up booty call, it is his sexual safety, not his sexual preference, that you should be most worried about.

Anyway, with Phillip being so young and still on his parents' health insurance, I insisted that he tell his parents about his HIV status so that he could get medical help. I've noticed that sometimes young people who get infected catastrophize their parents' reactions without giving them a chance to rise to the occasion. Sure, some parents are ignorant, some are angry, some are even violent, but most are simply devastated and need time to get over their sadness. Not only is their children's health threatened, but their chance at an easy life is diminished. It is possible to live with HIV, but life is never really easy again.

I also met Mike, a seventeen-year-old African American who was infected at birth—sometimes referred to as an "innocent victim." Here's more terminology I hate. Why should someone like me, who made an innocent choice, be thought of as deserving of HIV? That's what this dichotomy of the "innocent victim" versus the "guilty victim" implies. No one deserves HIV. Neither Mike nor I asked for this virus.

He didn't know life without HIV. His mother had passed away when he was a child, forcing him to bounce around from group home to group home. At times he sold his HIV medicine on the streets. It seemed like no stable household would accept him with his disease, but finally he was adopted by a very nice white family.

It didn't turn out to be *Diff'rent Strokes*. At the time of the conference, he was having a hard time dealing with disclosing his status to his new friends and neighbors. Instead he was constantly getting into trouble with the law and acting out in school. He asked me to his senior prom in Idaho but later changed his mind because I was so open and honest about my HIV status and he was not ready for the unwanted attention.

All of these young people—Maria, Phillip, Mike—helped me see that my life with HIV was just as complex, potentially rich, and joyful as life without it. They each defied the stereotype that life is not over once you're diagnosed; you still have your day-to-day problems and your day-to-day joyfulness. Instead they were each figuring out a way to live as fully as possible while still being realistic about their health. At Nashville CARES I had found a family and a sense of community, but being around Maria, Phillip, Mike, and all the other HIV-positive youth made me feel safe. The conference was a place of comfort, freedom from judgment. I took something valuable from each one of them, and I hope they took a little something from me too.

I realized that none of this would have been possible for me if I hadn't finally gone public with my story. Being exposed was sometimes scary—there was so much ignorance still in Nashville, especially among the black community. But the wonderful and unexpected consequence of my taking the risk was that I now had a community of youth just like me to depend on when things got rough. Community, I've learned over the years, is critical to staying healthy, with or without HIV.

There is a strange pattern in my life. Often at the moment that things are going really well—sometimes amazingly well—something terrible also happens. My first big experience of that was when I was getting my life together and then LaRena passed. My second was almost as painful.

Right around the time I received the award and was learning about this whole network of positive youth doing amazing work all over the nation, my mom broke the news to me that Lynn was dead. Ever the model of tact, she just handed me Lynn's obituary in a Waffle House parking lot on my lunch break. I was stunned into silence. All I could focus on was keeping my hash browns down. But when I returned to the office, I called her.

"Why didn't you tell me at the time, Momma?" I asked her, enraged.

"I didn't want to tell you right when she passed, because you had just gotten the promotion and then the award," she explained. "I wanted you to be able to have your moment of happiness."

I felt slightly betrayed, but I also understood and respected her reasoning. "When did she die? What happened?" I asked in disbelief.

"She passed away in December. You were aware that Lynn had been living with only one T-cell for a long time," said my mom vaguely. "It was her time."

It was crushing but, at the same time, fortifying. It made me feel like the award, becoming aware of this national movement of youth speaking out about HIV, had all been fated. I was meant to pick up where Lynn left off. I was determined to do her spirit justice.

"You know what, Momma? I'm going to keep going in her honor. I'm going to take this award and I'm going to really make it into something. You watch."

"I'm watching, Marvelyn. I'm watching, baby," she said. And if I wasn't mistaken, I could have sworn I heard something like pride in her voice.

On the *Price Is Right* stage at the
Kaiser/Viacom briefing.

chapter thirteen

The link between AIDS and death was no longer abstract for me. Not only had Lynn—the first HIV-positive mentor I had ever known—passed, but so had some of the young people I met at the youth conference, and people that I was working with through Nashville CARES started dying too. I was hanging obituaries on my wall and attending more funerals than I'd ever expected to.

Mourning someone who dies of AIDS is a strange experience. Everyone expects you to be ready—after all, that person had AIDS. But death always—even when expected—manages to come as a surprise. One moment a person is a beautiful, living, breathing force . . . the next he simply disappears. Since I'd never actually met Lynn in person, it felt even stranger—like my angel on the other end of the line had been disconnected.

Life got overwhelming pretty fast. I was getting better at all of the administrative tasks and serious leadership roles I was taking on at Nashville CARES, and none too soon. The infection rate in the Nashville area seemed to be rising every day, and the age of those coming in for tests was getting younger and younger thanks to our solid outreach among high school and college students. The word on the street was that Nashville CARES was the place to go, which made us a big success but also added to our stress.

As much as I liked having the guarantee of a little extra money at night, keeping up with my waitressing job was just too much.

Plus, that extra money wasn't exactly necessary. Most of it went straight to my shoe and handbag fund! (I lost a few friends and a goddaughter because of HIV, but I certainly didn't lose my fashion sense.) It was sad to say good-bye to all of my old buddies at the Olive Garden. That last night, I didn't even clock in. Instead, I left the managers a note, on an AIDS ribbon no less, and walked out in the beginning of my shift —with a sense of sadness but also destiny.

Letting go of waitressing felt like it really marked the end of an era. No more Shari. I was no longer ashamed of Marvelyn and what she had. No longer was I trying to get on my feet. They were firmly planted and now I was ready to fly.

The opportunity to soar came sooner than I ever could have imagined. It turned out that a representative from the Kaiser Family Foundation—a huge leader in the field of health education and promotion—had been at the Ryan White Youth Conference and heard me speak. She recommended to BET that I be recruited to participate in its National HIV Testing Day special.

I was thrilled, not just because it was BET, but because HIV testing means so much to me. I knew that prior to falling ill, I never would have voluntarily gotten an HIV test. I thank God every day that I found out that I had HIV in the hospital. As painful as it can be, it is always better to know your HIV status than not to know. Early detection leads to a longer and healthier life. The longer you wait, the longer that HIV is doing the crib walk on your immune system.

When the producers called and asked, "Would you be up for that?" I didn't have to think twice.

"Up for it?" I replied. "I got my MAC makeup ready! When do you get here?"

To celebrate the airing of the show, I threw a big viewing party in my apartment. The usual suspects were there: my sister Tab and her husband, along with Diamond, Jamiya, and Jamarius, and Cortney and her brother. I also invited a few coworkers and even some neighbors who wanted to know what all the fuss was about.

BET made a beautiful show—a snapshot of my life. It revealed the duality of living with HIV—that sometimes you are just a regular girl going to the mall, looking for something cute to wear, and sometimes you are unlike any of your friends, downing a handful of pills, struggling with nausea or uncertainty when you go out on a date for the first time and have to disclose your status.

I was honored that they showed my life in three dimensions. If more people got a glimpse inside the real worlds of those with HIV and AIDS, there would be far fewer overly optimistic people thinking that HIV is such a manageable disease with "all those new medicines." Yes, HIV is a manageable disease, but it is still hard as hell to live with. Take it from me: You don't want it. If you *do* have it, life goes on, but it's never easy.

Specials like this accomplish three things: They encourage people to get tested and face the truth; they reveal that, should you test positive, there is life beyond that fateful day; and finally, they emphasize that it is still critical that everyone take responsibility for preventing that day from becoming a reality.

I knew that the show was going to have an impact on my life and other people, but I couldn't have known to what extent. Immediately after the BET special aired, I started getting more attention than I could handle. The phone at work was ringing off the hook with requests from high schools, colleges, and, to my surprise, even churches throughout the South, all asking me to speak and make public appearances.

The high schools and college engagements were a thrill (teens seemed so hungry for the information I was providing), but I found the church circuit a bit more difficult. The questions were a lot tougher to answer. Some examples are "If you believe in God and His powers, why do you do medicine?" and "If you were a true Christian, why were you having premarital sex?" Not to mention that many of them requested that I not mention sex at all. "Just stick to an abstinence message, would you, Marvelyn?" a flamboyantly dressed Baptist preacher urged me.

I thought, "What in the devil am I going to talk about?" but nodded compliantly. I realized that opening the door to a conversation was the important thing.

Once I was in front of that pulpit, I knew there was only one man that I had to answer to, and that was God. I wasn't going to tell any lies in His house. I wasn't going to lie to my brothers and sisters. So, I preached the truth—with just a spoonful of sugar. No preacher ever stopped me. Instead, they embraced it and respected me for it.

I was starting to understand that, while I never wanted to compromise my values, it was important that I learn how to finesse the message for different audiences. I had nothing to lose, and the people—the young especially—had everything to gain from getting the real facts.

The BET show also demonstrated to the Kaiser Family Foundation what I was about—a mix of humor, straight talk, and wake-you-up facts. They were convinced that my story could reach a wider audience, and they were going to help me make that happen. Kaiser has a partnership with the Viacom, one of the largest global media companies in the world; Viacom owns Paramount Pictures, MTV, VH1, BET, Nickelodeon, and Comedy Central, among others media outlets.

To get started, Viacom invited me out to Hollywood to speak to a group of writers, producers, and directors about incorporating HIV awareness more fluidly and accurately into the media. I would have to miss work, but Patrick understood the significance of public education and media partnerships.

I won't lie. I had never, in all my life, been treated so good. When I got to Los Angeles, there was a limo waiting for me with Pepsi—my drink of choice—on ice inside. The driver took me straight to the Westin in Century City on the Avenue of the Stars, where I was given the most luxurious room I'd ever been in, much less gotten to stay in.

When I walked into the room, the first thing that caught my attention was the big white "heavenly bed" that the Westin is known for. I just stared for a second, scared to touch it because I didn't want to get the perfect white sheets dirty. I took two showers before I got into the bed. To give you a sense of how tight this place was, I was in the bathroom taking pictures of the toilet. There was a note instructing me to eat whatever I wanted, drink whatever I wanted, to generally make myself at home. They obviously didn't know what "home" meant to me.

I soaked it up, taking advantage of every luxury and indulgence. I ordered *Herbie Fully Loaded* on the TV and a burger and some fries, not fully understanding the room service menu. It was the best burgers and fries I had ever had in my life, and I must say that the plate design was nice too. How do you take a burger and fries and make it into a plate of art?

I had to remember that this trip was not about soaking up the L.A. sun and going to the Ivy for lunch to stargaze; the Kaiser family had me working. The next day I headed to CBS Studios. I met Tina Hoff, vice president and director of the Entertainment

Media Partnerships at the Kaiser Family Foundation, and she gave me the rundown of the day, clueing me in on the major players at the networks.

When I walked into the conference room, I have to admit, I couldn't distinguish one from the other. They all just looked like a bunch of cold suits. I was intimidated. I could sense that some of them were irritated that they had to be there, taking time out of their busy schedules. But the self-loving Marvelyn took over and told them my story. I painted a picture of my hometown, the chicken-wing spots, the talented but lost teens, the rampant hopelessness and unprotected sex, my shocking diagnosis, loneliness, depression, and all the growth and change that came afterward. I told them about my commitment to prevent other young people—especially young African American youth—from contracting HIV. I told them about my dream of living a long, exciting life, and I did it all with a smile on my face and lots of humor.

The result? The crowd was either laughing so hard that they cried, or laughing through their tears of sadness and empathy. The perfect combination. All of a sudden, those professionals seemed far more human.

To my surprise, I learned something else during that meeting—that speaking just seemed to come naturally to me. It wasn't something I had to train to do or prepare for. I definitely did get nervous, but I guess I'd always been dramatic, always been a performer, so this was a natural progression. I was no longer ashamed of my story because I saw the way it moved people. It was a joy to finally be able to talk about my status, especially with such powerful people, and realize that they thought I was strong, not a leper.

These people knew no more than the folks back home about HIV and AIDS transmission, but they were willing to be educated.

Initially they may have questioned why they had to go listen to some nobody from Nashville talk about AIDS, but Viacom did make it a priority. It was always a respectful space, and when it came down to it, it was a pretty welcoming crowd—one that was now informed.

I would later find out that because of that meeting, UPN's *Half & Half* writers designed a subplot based on my story. Michelle Williams of Destiny's Child would be playing my character. Needless to say, I was flattered that my talk had inspired this fictional but educational plot for one of UPN's most popular shows, but I won't pretend I didn't think, "Damn, I could have played myself. Or they could have gotten Beyoncé . . ."

A week later, in Nashville, my phone rang again. It was a producer from *America's Next Top Model*. The glow hadn't even worn off from my trip, and now Tyra Banks wanted me to be on her hit show, teaching the models how to do a public-service announcement. On the inside I was screaming, but I tried my damnedest to play it cool as I spoke on the phone with one of Tyra's producers.

A sense of panic took over my body once I hung up the phone. Patrick had let me go to Hollywood the first time because we'd both agreed that it was important to do so. I was scared he would think, "Enough already, Marvelyn." People in the office were a little ticked. After all, when I wasn't around, my coworkers had to pick up the slack. I could understand where they were coming from, but what was I supposed to do?

Although I wasn't officially working for the Kaiser Family Foundation, they were quickly becoming foundational to my ability to reach out and help people on a large scale. The Kaiser family believed in the power of my story, even when I wasn't sure how the television audience would receive it. I still didn't know what it could do, who it could save, but Kaiser saw into the future and knew that I, like so many people, had the potential to make a difference. Nashville CARES, on the other hand, was important as well—it made a huge impact on a local level. So, I found myself having to make a decision: gaining access to a national and even global audience by working more with Kaiser, or staying in Nashville, continuing to educate and help people in my home state (which, despite all its difficulties, was a place I loved very much—it was home).

"Patrick, you got a minute?" I asked as I tentatively peeked into his office.

"Of course, I've got a minute for all my workers," he said.

"You won't believe who just called," I said, settling into the chair across from his desk. "*America's Next Top Model*! They want me to go back to Hollywood and tape an episode where I teach the models how to do a public-service announcement. I know I just missed work, but this can't last," I said, getting it all out so I could breathe again.

"It hasn't been easy around here picking up your slack, and some people are getting a little irritated by it, but I know how much people long for and need this information. When you get back you're going to have to work extra hard to catch up. I'm so proud of you, girl!"

I was, once again, reminded that Patrick was a true mentor. He wanted to see me thrive within Nashville CARES, but even more, he wanted to see me reach my highest potential.

I flew back to Los Angeles the following day. The taping of the show was in the Hollywood Hills, a neighborhood I'd seen on *Entertainment Tonight* but never actually dreamed I'd hang out in one day. The producers instructed me to just sit there when the models walked in and not give anything away about who I was or why I was there. They all clamored on a couple of couches, and I heard them whispering to one another, "Is that Iman's sister? Does Tyra have a sister?" I was so flattered, but I tried to keep as straight a face as I could.

Then we got down to business—I told them what I was there for and let them know that I now had the honor of directing who-ever won the challenge the previous day through a public-service announcement for the Kaiser Family Foundation and Viacom's kNOw HIV/AIDS campaign—to be aired on Viacom's various networks. Ultimately, the winner was Furonda, and she chose her friend Nnenna to help her out. They squealed with excitement, while the other models seemed disappointed.

The stereotype is that models are dumb and snobby, but those girls were the exact opposite. I hung out with them all day, chat-ting, listening to music, and joking around. I told them that if I was a little taller, I would have given them a run for their money on that show, which made them laugh. The tension that had filled the room quickly faded as I told them my story and answered their questions. Even the one who was supposed to be the bitch of the show, Jade, was crying when she heard what I had been through. Danielle, the winner of the season, stated that she did not know anybody personally who had the virus.

I hear that a lot, but according to the CDC, each and every person in the United States knows someone personally who is living with HIV or AIDS. There are just too many of us infected for anyone to escape the disease entirely. Think about it like this—you may not know that you know someone with HIV. So many people don't "come out"—even to this day—because they are afraid of the backlash that might occur. There is still so much discrimination in employment situations and enough ignorance among the general population to cause family members or friends to abandon their loved ones when they learn the news. You might not know because someone is scared to tell you. You might not know because the person who is infected might not know. You might not even know about your own status, because you haven't taken your risk seriously.

I was finding that people wanted to have these conversations in the most unlikely of places. During downtime on the show, cameramen kept asking me for information. They wanted to know how they could get tested (*Is it painful?* No. *Is it accurate?* Yes.) and what HIV really meant in terms of long-term health effects and quality of life. It was a growing trend—the people I least expected to approach me needed to know what I knew.

When the episode finally aired I was only on the show for four minutes, but considering what I took away and the people whose lives were changed that day, it had a lot of impact. I took a leadership role on a major national show. I held my composure in front of the camera, explained my presence there articulately, opened up some of the models' minds, and even helped a few cameramen. I walked away with even more confidence—and a lifelong friend, Furonda, the challenge winner.

The domino effect was in full force: I did an MTV news special about stigma and discrimination, a cover story for *POZ* (a magazine for people living with HIV/AIDS), a spread in *Real Health: The Black Wellness Magazine* (*POZ*'s sister magazine), and another for *HIV Plus*. Then *The Tyra Banks Show* called. Tyra wanted to meet me (she hadn't been on the set the day that I did my taping with the models) and talk about living with HIV/AIDS and the increasing infection rate among young black women in particular.

This was really the population I was most invested in speaking to. According to the CDC, the rate of diagnosis for black women was nearly twenty-three times the rate for white women, and the primary way that so many black women were getting infected was "high-risk" (i.e., unprotected) heterosexual sex. That was me. That was my people.

It was so exciting to sit next to someone I had admired for years. I had actually written Tyra a letter once, years earlier, telling her how much I envied her strength and courage. Just like anyone else, I guessed, she'd been faced with her share of challenges, but she handled it all with such grace, style, and class. Plus, she is an amazing businesswoman, and I was starting to admire those qualities more and more.

In the middle of the show, she reached over and gave me this huge hug, basically saying with her body language, "Look, America, I can hug her. I can touch her." It meant the world to me. Even with my newfound stardom, I was still dealing with people not wanting to be physically near me sometimes, worried about touching me or eating the same food I was eating. Tyra sent a message to the entire nation that HIV was not contagious by touch alone. She sent the message that those of us who are positive still deserved love and affection.

Don't get it twisted—life wasn't all media hype and glamour. Around the time *The Tyra Banks Show* aired, I came home one day and found "Fagfucker" written on my door. I noticed that when I got into the pool at my apartment complex, some of the mothers would grab their kids and pull them out of the water. Despite my so-called fame and my newfound beauty, at this moment I was reminded that I had still had one of the most stigmatized diseases in the world.

I was reminded yet again when a family member—someone I loved—died, reportedly of cancer. Or at least that's what it said on her death certificate. I knew otherwise, because she was a client of Nashville CARES and I ran into her one day in the hallway. That's right. Someone in my own family had AIDS, and no one knew it. Not her mother or her children. To this day they don't know, and they never will. People die of complications from AIDS, not AIDS itself, so it is fairly easy to mask AIDS status even in death.

There was something about her passing—passing without ever owning up to her true self—that broke something inside of me. It felt like I was reexperiencing the loneliness I felt following my diagnosis. I was sensitive. I was vulnerable. I was still, after all this time, alone.

Everyone around town knew "Marvelyn Brown has HIV," but now—thanks to all of these television opportunities and outreach programs—they had a face to go with the name. They talked about me in clothing stores, whispered and pointed, sometimes even asked for my autograph. But instead of making me feel less alone, all this attention just made me feel more exposed. To those people, I was an idea now, a character, not a real, flesh-and-blood person. They were kinder to me not because of my soul but because I was on national TV.

And the worst part was, I couldn't yet live up to the image I presented on TV. It was as if I was an actor playing the role of my life. I would stand on stages in front of one thousand eager faces and urge everyone to protect themselves, get tested, and no matter what the results, to believe in the possibility of a full, happy life, but I had yet to find my own inner peace. I'd tasted self-love, but it was still elusive.

Another unintended side effect of my fame was that I couldn't keep up with my responsibilities at Nashville CARES. I was falling behind and felt that my community was suffering because of it. Finally, Patrick and I sat down and had a heart-to-heart. We both knew what had to be done. Though I would always have a special place in my heart for the organization, Nashville CARES deserved someone who could devote full energy to the work on the ground. I left on good terms, giving Patrick a huge hug before I walked out the door. It was a scary feeling, but it also felt like the right thing.

Without a full-time job, I was truly on my own again. Now that I had all the time in the world on my hands, my availability was no longer a problem, but my safety certainly was. Being a semipublic figure had increased the hate mail and threats I got—through MySpace, notes slipped under my door, and the like. Even if it was all hype, I felt like I needed a community around me.

Fame was a double-edged sword for me. On the one hand, it opened the doors I needed opened in order to educate the public, to really reach out to Americans and spread my message. Television was such an amazing way to reach youth who wouldn't go to an AIDS program voluntarily or show up at a community-center function on safe sex. They could just be lazing in front of the TV after school, watching BET or *America's Next Top Model,* and learn the truth about HIV and their vulnerability to it. That made me so happy.

It was also really fun work. Who wouldn't want to meet celebrities and have the chance to speak publicly about something they care so deeply about and experience so personally? It was a dream come true to stay in fancy hotels, chat with famous, brilliant people, and get my face out there.

But it wasn't all limo rides and celebrities. It was also exposure—having everyone, even perfect strangers, know your business. It was unsafe. The world is still filled with so many misconceptions about HIV and AIDS and so much discrimination. I needed my family and close friends more than ever. I needed their support, but even more, I needed their protection from all of these outside forces that were threatening to sneak into my personal life—a truly discomfiting feeling.

And it was a lot of pressure. I was starting to understand that, as they say, "with power comes responsibility." I wanted to live up to the opportunities that God was giving me, but I was still young and confused. My first mentor, Lynn, had died, and my second, Patrick, wouldn't be as accessible now that I was leaving the organization. How was I going to stay strong and choose the right path? Who was going to teach me how to juggle all the demands of being a public spokesperson?

And where was I going to go to feel safe? My mom was constantly traveling, so her house wasn't my first option. The perfect place, the place I wanted to be more than anywhere else, was in my sister's loud house, packed with naughty nieces and nephew, fun, and family. That sounded like heaven right about then, even though I knew that I was going to be jealous of her beautiful family.

I showed up on her doorstep, and when she answered the door, I just looked her in the eyes and said, "Sis, I need you."

She didn't say a word, just opened the door and motioned for me to come on in.

My daddy. **Photograph courtesy of Fred Cowan**

chapter fourteen

Now that I was back living in a safe situation, I felt my confidence return. Sha, Kendria, and I started to go out again like the good old times. I hoped that my television appearances might make disclosing my status a bit easier—local guys were more likely to know about my status through my unique name. But not having been out to club with the girls in a while, I was still nervous. At first, I stood back from the dance floor, leaning against the wall and watching people have fun. Sha came over and shouted over the music, "Marvelyn, why aren't you dancing? You ain't the same no more. You used to be the life of the party."

I shot her a look and said, "I got HIV, Sha."

"Girl, I know that. Everyone in town knows that. You can still have fun!"

I appreciated her encouragement, but I also didn't think she understood how much I'd been burned since being diagnosed. It wasn't easy being young and in the dating world with a highly publicized disease. Everywhere I'd turned, there had been heartache and humiliation. As I leaned against that wall, I thought about some of what I'd been through in the last couple of years with men.

The first night I went to the club after my diagnosis, all I wanted was to be a normal nineteen-year-old again. I'd been sick and in the hospital for so long, then sitting in my car looking at the sky and feeling sorry for myself for even longer, and now I just wanted to be out and about. I wanted to feel young, healthy, and beautiful again. I wanted to dance. I wanted to have some fun.

But while I was dancing with this dude, his homeboy gave me the eyes and whispered something in his ear, and my fantasy ended. The jerk literally picked up a drink and poured it in my face. "Stay away from me, girl. I know what you got!" he yelled over the sound of his friend's haunting laughter. I went home crying and stayed away from clubs and the public for a while.

My year at the community college was almost nothing but hell—the finger pointing, the name calling, the out-and-out rejection. But I write "almost" because there was one thing that kept me showing up to my remedial basic math class: a cute guy, of course. I got such secret joy out of looking at him and his handsome face and thinking that one day I would have a husband who looked as good as him, despite how my life was falling apart all around me. I'm sure he never noticed me, because I walked around looking like a sad black cloud. Besides, everyone knew I was the token girl with the deadly virus.

But one night, speed walking to my car after eating at the Hard Rock Cafe in downtown Nashville to celebrate my birthday, perked up with a little makeup, tight jeans to show off my Coca-Cola shape, and some killer knee-high boots, I realized that he was driving the car that was following me. "Hey, sexy lady, I don't want to waste any of your time or my gas. Can I get your number?" he asked.

"Sure," I said, giving him a coy smile. But as he was typing the number into his phone, I couldn't help saying something: "You know, we had a class together last semester at Vol State, right?" It took a few seconds to register, and then I saw the sweetness drain right out of him as he realized that I was "the girl with AIDS."

His texting slowed down as he finished putting in the number, and then he just drove off. Nothing so much as a text message followed. Knowing that I could have had him without HIV gave me temporary satisfaction, but the rejection cut deeper. All he saw was the virus, not the girl who had it.

Then there was that one guy who I dated who said he was "cool" with my status but just didn't want to "dwell on it." Translation: Let's both pretend this never happened. Any time I bought up the stresses of having HIV, my medications, my visits to the doctor, he would urge me not to complain so much and stop acting like a little-bitty baby. His theory was that talking about anything difficult only made it worse, which led to a whole lot of pretending on both of our parts.

When I finally realized that I was tired of tiptoeing around and playing a part, I quit his ass. I didn't like the feeling of being alone, but I realized that having a partner who didn't support me or demand honesty was no better than being alone. When I broke it off, he had the nerve to say, "You should be happy that somebody wants to talk to you. Remember, you *do* have HIV."

I looked him dead in the eye and said, "I do remember. It's you who doesn't seem capable of dealing with real life. If it wasn't for HIV, you wouldn't even be here right now. You should be saluting it!"

"What's that supposed to mean?" he asked, anger scrunching up his face.

"That for a minute there, I thought that I had to lower my standards because I had HIV. I wouldn't date your ass without HIV, and I shouldn't be dating you now. I deserve real love, not this half-assed pretend game."

Remembering my strength at that moment got me up off that wall and headed to the dance floor. Sha was right. I deserved a good time. Plus, I hated to waste a cute outfit feeling sorry for myself.

With the Ryan White Youth Conference approaching again, I knew that I would have another opportunity to be among *my* people, feeling totally comfortable. Being around HIV-positive youth gave me a sense of security and nourished me on my journey toward self-love. I needed that so much.

When I walked into the conference room, I scanned the crowd, looking for old friends, but a new face caught my eye. He was a little older, light-skinned, cocky, really cute, just they way I like them. He came over and we chatted, immediately hitting it off. We'd both been involved in planning the conference, so we'd actually spoken on the phone, but I'd had no idea he would be so fine, and evidently he wasn't disappointed with my looks either.

"You must be Tim," I said.

"And you must be Marvelyn," he replied. "I didn't really know what I was dealing with on that telephone," he continued, moving closer.

"Well, apparently neither did I," I said, smiling slyly.

We hung out all that day, talking about our lives, our twin struggles with HIV, our futures. I was feeling brave but not that brave, so I wrote him a note. I handed him the folded paper with a heart on it and walked away. You would have thought that I was sixteen again, but I was fully twenty-one. I hid behind a door, peeking through the cracks to watch him read it. Suddenly I felt silly. I wasn't a teenager. I didn't want to be scared. So I walked out from behind that door and right up to him and said, "I'm catching feelings for you."

"I know, Marvelyn. When I first heard I would be working with Marvelyn Brown, I thought that you were stuck-up and would have an attitude. I can't lie I seen you as direct competition too. But now I see you take pills and put up with stigma like the rest of us. I'm already hooked on you too. But I have to tell you something . . ."

I laughed. "Boy, I know you're positive. What else could you possibly tell me?"

He laughed, then said, "Naw, I have to tell you that I have a girlfriend."

If there had been a sound track for that moment, it would have been screeching brakes. "A girlfriend?"

"Yeah, we've been dating for a while. She's actually negative, so I don't really feel like she gets me in the way you do. We haven't even had sex yet, but it does feel good to know that she accepts me even though I'm positive. I think I get sucked in by that."

He had been HIV-positive almost his entire life because of an infected blood transfusion, so I imagine he dealt with even more stigma than I had in my short years of being positive. "But Marvelyn," he went on, "the feelings are mutual. If I didn't have a girlfriend—"

"But you do," I said, not about to mess with someone else's man.

After the conference ended, we kept in touch, talking almost every night on the phone. We'd chat about anything and everything until his girl was around. He admitted to me that he only kept his girlfriend around because it made his momma happy. "Another frickin' momma's boy?" I thought, exasperated. It pleased her that he managed to have a relationship with someone who was negative, as I would later find out, an all-too-common sentiment among relatives of HIV-positive people.

We also sent little e-mails back and forth, expressing our feelings for each other through pictures and kind words. Once I even sent him a package of chocolates at his workplace. We weren't calling each other boyfriend and girlfriend, but we were doing all of the things that defined that kind of a relationship.

Then the opportunity came up for us to actually spend some time together. One of our mutual contacts was doing an HIV youth tour around the country in an RV. We would both be on it, at least for one week, over the spring. I counted the days, putting big X's on my calendar.

The only rule on the tour was that speakers could not have "relations" (i.e., sex) with one another. We broke that rule every single night that he was there. We did it in Texas. We did it in Kansas. Even in Iowa. We did it in just about every state in the Midwest. Once again, I let my passions get the best of me.

Even though I was positive, being with a positive guy scared the hell out of me at first. I knew the risks. If two infected people have unprotected sex, they can reinfect each other and multiply the viruses in their bodies. This amplification of the virus can push someone from HIV status to having full-blown AIDS, or even create resistance to various medications. But here we were, preaching safe sex all over the country. We could certainly manage to take precautions ourselves.

I think part of my fear was just psychological. Though I craved the familiarity, I didn't like the idea of dating someone who dealt with all the same things I did on a daily basis. It just didn't seem romantic to have a husband with whom you constantly discussed meds, doctor visits, and T-cells. It was bad enough that I needed my health to be front and center in any relationship. To have two positive people seemed overwhelming.

Plus, the idea of always worrying if he was taking his meds or going to his doctor regularly and the thought of death were so painful. I didn't want to worry. If I was going to worry, I wanted to have the same worries that the average girl has. *Did he pay the electricity bill? Did his team win* Monday Night Football *so he won't have an attitude?* I didn't want to deal with AIDS-related issues all the time.

But on the flip side, it was good to have someone who knew what it felt like to be judged for a disease that you have no control over, who knew what if felt like to have diarrhea day after day and the tricks for dealing with it, like how to wipe the seat good in the public bathroom. It felt good not to hide my medicine in fear that my horse pills would scare the guy off. And oh, not to mention, it was good to have sex with a male who was not checking to see if the condom broke every five seconds. It felt damn good to have someone who understood me and some of the stuff that I was going through. Honestly, I just thought that it would be easier.

By the end of the tour, he looked me straight in the eyes and said, "I love you." I couldn't believe he said it first. I was so used to being the one to say it and then waiting for months before hearing it back. We had such a deep connection when we were together that somehow neither of us could keep our other lives—his girl-friend, our geographical distance—in our minds. Occasionally we would discuss his girlfriend back home, and he would complain, talking about how things weren't really working out, how he was just waiting for the right time to break it off. I ate it up like those girls in the Lifetime movies. I thought that he would leave her. She didn't understand his hurt and his pain. I did.

Clearly I was caught up in his spell, and still—after all these years—not listening to my wiser self. I let the lust and my craving for comfort get in the way of what I knew to be true—that he wasn't going to leave his girlfriend and, more important, he was never going to be a substitute for my own self-love.

As soon as he got home, I got the phone call: "Marvelyn, I can't be with you." I could almost hear his momma and girlfriend breathing at the other end of the line.

"Why not?" I asked, totally thrown off but trying not to act desperate.

"My momma would never approve of me being with an HIV-positive girl—especially with someone who is in the spotlight and is not ashamed of her status. It would never work out. I have an HIV-negative woman who loves me. Think about it. She has to care. I can't let that go. Come on, Marv, you know how it is." I could picture him, his eyes filling with tears. He sounded totally powerless but also completely resigned to his fate.

Hearing that sad quality in his voice helped to lessen my attraction for him. As for his momma . . . words could not express how complicated my feelings toward her are, but I will try. You see, my HIV-positive love contracted the virus at the young age of one. His mother worked hard to keep their little secret from family, friends, and the school system. She knew the stigma and discrimination that her son could face, and she wanted him to live as "normal" a life as possible.

So, instead, her son harbored deep shame. He told no one about his status, including childhood homeboys, girlfriends, and sexual partners. I understood why his mother thought she was protecting him, but I didn't understand why she was hurting him at the same time.

Now that her son was ready to take his story public and own up to the mistakes that he made, she was still preaching silence.

He could have infected many people and many lives through her selfish influence. But I was not about to let either of them affect mine anymore. "Do what you got to do. I'm steady doing me," I said, and then hung up the phone.

When I got back from the tour—a little heartbroken, I have to admit—I stayed at Tab's house and soaked up the love of all my nieces and my nephew. But I wasn't much for leaving the house. Everywhere I went people were recognizing me, and as much as I love attention, it got to be annoying pretty fast. The phone was ringing off the hook with requests from coast to coast for me to speak here, appear there. I was getting recognized on the street: *Oh my God, aren't you that girl with AIDS? Excuse me, but were you on BET? Don't I know you?* It was getting completely outrageous. I couldn't even pick up some Pepsi without someone hounding me. *Are you allowed to drink that?*

I began struggling between my identities as "Marvelyn Brown, the AIDS activist" and "Marv, the homegirl." HIV was becoming a 24/7 issue, and although I loved to educate people and drive home the message of HIV prevention and education, I did not know who Marv was.

On my birthday—May 7, 2006—I decided to give myself a present: a new start. I bought a plane ticket, with some of my sky miles, to Washington, D.C., where one of my new HIV-positive friends from the conference was living and had offered to let me stay until I figured out my own situation. D.C. seemed like the perfect place to find myself without sticking out too much—after all, that city has ten times the infection rate of other American cities, and black folks are disproportionately affected. Some studies estimate that as many as one in twenty young black Washingtonians has HIV/AIDS. I would be in good company.

As the plane landed in that flatland of D.C., I had that familiar feeling of being back on my own, starting from scratch. I wasn't living in my car, thank God. I wasn't facing the initial sting of stigma and humiliation, thank God. But I was starting over. Again.

It wasn't just my heart that was broken. About this time, my health was in real jeopardy as well. Without a full-time job, I lost not only the security of a steady paycheck but the ability to pay for health insurance too. With my medication costing as much as twenty-five hundred dollars a month, there was no way I could afford it all on my own. This wasn't a matter of cute shoes and bags. I needed those medicines to live.

I could almost feel the virus attacking my immune system and my T-cells decreasing by the millisecond. After missing my HIV medicine for a week, I called Dr. Laya in a panic. She coordinated with a local clinic in Tennessee to send me some meds through a special program they had for HIV-positive clients who could not afford their meds or insurance. The medicine came from either patients who switched their regimens and, as a result, had leftover pills or patients who had died and their families turned in their meds.

The generosity of others saved my life. I was so grateful for that, but I was also dumbfounded by the way the real world worked. I had been on *The Tyra Banks Show*, MTV, *America's Next Top Model*, BET, and in countless magazines telling the whole world that they should get tested for HIV/AIDS and make sure they got into early treatment, that they should be positive. I was, just that month, featured in *Newsweek* under the headline "The New Activist, Marvelyn Brown, 22," and the write-up claimed, "Now, healthy and on medication, she teaches young people to protect themselves." It should have read: "Now, temporarily healthy and unable to afford medicines, she is struggling in D.C."

I was trying to maintain my newfound HIV/AIDS spokesperson image, but in truth I was facing as many problems as some of those I was trying to help. Contracting HIV from someone I generally trusted and cared for made it hard for me to trust anybody—whether it was friends or family or whomever. I had a lot of guilt. I would repeat the same questions over and over in my head: Why didn't I insist on a condom that day? Why did I have to be so damn vulnerable? What if?

Then there was the regret. I thought of the men who had approached me pre-HIV whom I'd just snubbed. I had the painful realization that, not only was I ashamed of myself for having HIV, but I was ashamed of the person that I was before HIV too. I had never been emotionally stable. I had been living the life—going to school, hanging out at parties, enjoying popularity, but I'd always lacked self-love.

It was strange to look back. I'd been surrounded by people who wanted to be in my circle of friends, people who loved me. How could *I* not love me? If I had loved me, I wouldn't have given someone else the power to determine my future by having unprotected sex. I was facing some of these truths for the first time in my life. It hurt, but I knew it was necessary.

Around this time, I was touched by a call I received from my father. He would contact me from time to time, just like when I was a little girl. After I told my story in the Nashville papers, I didn't hear from him for ages. But in this phone call he explained, "Marvelyn, when you were first going public, I didn't agree with it."

"I figured as much. I'm sorry if it hurt you, but it was something I had to do," I explained. My soft spot for my dad had never gone away.

He continued, "No, but I called because I just want to say I'm proud of you and what you've done."

"Thank you, Daddy," I said. At that time it was the last of the medicine that I needed to keep me going.

It was moments like these when I realized that my relationship with my father—or lack thereof—was one of the sources of my difficulty with men. It wasn't just the HIV that stood in my way. It was also the unresolved business of my father's missing love.

You see, when a girl grows up without her father, or just the occasional, unsatisfying experience of him, there is a pinprick of a hole that she hides inside. I think that hole grows bigger and bigger if she ignores it, especially as she gets older. She tries to fill it with substitutes—boys who boss her around, men she doesn't feel worthy of but wants nonetheless; sometimes she even has a baby, thinking it will finally numb her pain.

But none of this ever ends up solving the initial problem, which is the innate ache for the father who loves her. Some—though not all—of my struggle to love myself was also a struggle to understand how my father didn't love me enough to stick around. I know so many young women who can relate. Deep down, we have a hard time believing we are worthy of love when our own fathers—the first men we ever knew—show us otherwise.

And this is the worst part. It is not their words but their actions that betray their inability to love right. My father wasn't around much, but when he was, he expressed his love for me fairly freely (more freely than my mother was ever able to). He was a loving, kind man. But sometimes I wondered if that made his absence even harder on me. The gap between his words and his actions was so wide that I just fell right through. It would be a long climb back up to self-love, and I couldn't depend on any other man to give me a hand up. I would need to take those steps on my own . . . even if my stilettos got scratched on the way up.

On the red carpet at the BET Awards.

chapter fifteen

My favorite Bible verse is Hebrews 13:5: "Never will I leave you, never will I forsake you." I had read it faithfully from the time I was a little girl, getting such comfort from the idea that God would accept me and offer me His love even I didn't accept myself. But I was also starting to understand that God's love of me wasn't a substitute for my love for myself. He wouldn't forsake me. I needed to stop forsaking me.

I decided that I wasn't going to play the fool anymore, that I wasn't going to rely on men as substitutes for self-love, that I wasn't going to be financially dependent on anyone or compromise my health. I was set on getting my life together, once and for all. People would often write me emails and tell me, "You are my hero." Well, I wanted to be my own hero.

As a strong believer in God, I believed He saw how resolute I was in turning things around, and my life started to change. I got a job offer from *POZ*'s brand-new editor-in-chief. She wanted me to move to New York City (the city of dreams!) and take over as an ambassador and community outreach coordinator for the magazine. I was done blocking my own blessings. I would not be afraid.

Right after I accepted the job at POZ, I got word that MTV—having seen me at one of the Kaiser Family Foundation briefings—wanted me to shoot a series of eight public service announcements on HIV/AIDS with award-winning director Joel Schumacher. At first, I have to admit, I wasn't sure who Joel was, although the last name definitely rang a bell. I Googled him and soon found out just who I'd be dealing with—the director of *Batman & Robin, A Time to Kill,* and *The Phantom of the Opera,* among other notable films. Working with Mr. Schumacher sounded like a huge responsibility—but also a wild opportunity. I knew that the part of me that always rises to the occasion would kick in once I stepped into the studio. But I didn't know how powerful that particular message was until it was nominated for, and then won, an Emmy.

And if that was hard to take in, I then got a call from BET, letting me know that I'd just been named one of the Top 25 Heroes of the twenty-five years of the AIDS epidemic, along with people like Alicia Keys, Magic Johnson, and longtime HIV and AIDS activist Phill Wilson. We were asked to film some commercials and be honored in an awards dinner at the International AIDS Conference in Toronto, but even more exciting, we were all invited to the BET Awards, coming up shortly in Los Angeles, and to walk the red carpet.

I arrived at the red carpet in my Kenneth Cole get up, confident my make-up looked stunning but afraid that the sweat from nerves and the L.A. heat would mess it up. I looked around for my model escort. She was obviously fashionably late. There were cameras flashing everywhere, film crews, two long ropes that kept out the throngs of "ordinary people" who had shown up to catch a glimpse of the celebrities heading in. I couldn't get over the shock that my ordinary non-singing self was on the inside of the rope.

I saw Vikki, a friend of mine from the BET Rap-It-Up family, and immediately latched on to her but not for long because she had an accessory, a walkie-talkie because she was working. I managed to walk around and observe and even rub shoulders with a few celebrities. I said "What's up"—trying not to act like a groupie—to Jermaine Dupree, Janet Jackson, and their million body guards. I even saw Kelly Rowland and Michelle Williams, but I swear I got too nervous to say more than a word to them. (Apparently Beyoncé was backstage already, getting ready to open the show.) But I did manage to smile at Nas and Kelis, and I swear she smiled back.

After about an hour parading around that red carpet, we finally went in and took our seats. Furonda, my model date, and I managed to get ourselves into the fifth row. Neo was directly in front of us. Rihanna was to the left of us. Mo'Nique—the lady I admire and love—was right beside us, and Diddy and his team of kids were a few rows up.

I could feel it in my heart—God was not far behind me.

The next call I received, was the polar opposite of good news. It was my sister Tab. "Hi, girl," I said, casual as always. Tab and I talked on the phone quite a bit since I moved to Washington, D.C. It was our way of staying close and her way of knowing that her baby sister was okay in the new city. She'd call me about any old thing—telling me Jamarius's "joke of the day," describing Jamiya's new dance move, or going on about how many points Diamond hit in her basketball game. Sometimes she just told me all the beauty shop gossip. But this obviously wasn't one of those calls. She was sobbing so loudly and screaming so hysterically that I couldn't even hear her response. "Tab, calm down. Girl, calm down, I can't even hear what you're saying," I pleaded with her.

"Dad . . . Dad is . . . hospital . . . I don't know what to do."

She couldn't complete a sentence, but I got the gist of what was going on. My father was in the hospital. It was serious. When Tab was finally able to catch her breath she said, "Marvelyn, they don't think he's going to make it."

From what she'd pieced together from our aunt, my dad had been found passed out after drinking some moonshine that may have been poisoned with antifreeze. I'd always known that one of my aunties was a crackhead; hell, I'd seen her do drugs right in front of me. She is, thank God, now clean. But I never really suspected that my own father had a drug and alcohol problem. People rarely resort to drinking moonshine unless they've got a real addiction. I guess this was confirmation. Growing up, the family had mostly shielded me from the truth—my daddy was a crackhead too. I still don't want to admit that.

Ever since being diagnosed with HIV, I'd always thought I would die before my daddy. I know that's not nature's way—the parents are never supposed to bury the children—but I'd just sort of figured it would end up that way. It was a shock to find myself sitting at my father's funeral back in Nashville.

The service was short and to the point. To my disappointment, our half siblings (whom I'd never met) didn't show up, but their names, my half-brother Dorian and my half-sister Michelle were mentioned in the obituary. No one really spoke about his life, but one of my cousins sang a beautiful song for him. My mom was there, with huge black Ray Charles sunglasses on; I'm sure she didn't want anyone to know how hurt she was over his passing. They had so many unresolved feelings for one another, so much anger, so much disappointment, and still, to that day, so much unexpressed love.

When my father was alive, I hungered to spend more time with him, hear from him more often, know that—though he had left us—he still truly loved us. I was surrounded by women my entire life, my mom, my aunties, my sisters. I missed having a man around. So when he died, the feeling of missing him was familiar, just a little deeper, a little more final.

I went up to the open casket and stood there a moment, carefully laying a photograph of his girls—me, Mone't, Tab—inside, saying a prayer. I can't remember what I prayed. I'm too honest to make it up. But I can tell you that I pray for my daddy every day. I ask that God give him a beautiful resting place in heaven, somewhere free of addiction, pain, and regrets, a land where he can finally love as good as he always meant to. We will meet again.

The day of my daddy's funeral, I went out and got the name of a guy I had just started dating tattooed on the small of my back. As always, my supportive friend, Cortney, was there beside me. She didn't agree with me one bit, but she knew the stubborn Taurus in me was not going to budge. I can still remember how that needle felt piercing my skin, the most welcome distraction in the world from the emotional pain I was in that day.

Of course, when he saw it, he got scared half to death (scared him right off, truth be told), but I think it scared me even more. I'd always gotten tattoos to mark big events in my life. I had one to commemorate my sixteenth birthday and a tiny red AIDS ribbon on my wrist shortly after my diagnosis. This last one, I convinced myself, was to mark my Dad's passing, but by putting a guy's name on it instead of my father's, I realized, was talking the self-love game, but had not fully absorbed it.

My father's funeral was on July 1, 2006. On July 4, I got on a flight to New York City to start my new life. Yet again, I knew that there was a better future for me and that I had to believe in that possibility, no matter how scary it might be. I had only one carry-on bag, a purse, and a thousand dollars to my name—having spent the rest of it on getting home to Nashville for the funeral—and no clue where I was going to live once I got to the Big Apple.

I stepped out into the hot, humid New York City air and there were no limos to pick me up and whisk me away to some fancy hotel. This time I had to take an overpriced yellow cab that smelled like last night's party to my new office, where I arrived exhausted, scared, and broke. This moment was not like those times before.

When my taxi pulled into midtown, I can't say I was initially impressed. Everything seemed dirty and overcrowded. But as soon as I stepped onto Broadway and looked up at the neon flashing lights of Times Square and the people rushing all around me, snaking in and out of traffic, I knew I was in the right place.

New York struck me as a place where I could just blend in. Everyone seemed focused. Everyone had a mission. People in New York are going places, literally and metaphorically. I realized that with this much anonymity and ambition, no one would give a damn that I was HIV-positive. This was a place where I could find my own identity, without a lot of haters or groupies. And that's just what I was aching to do at this point in my life. I wanted to learn Marvelyn.

I spent those first few days at my job just trying to find a place to live. All of the stereotypes I'd heard about New York City seemed to be true. It was dirty and unfriendly. People everywhere were looking to make a buck. Upon answering one Craig's List ad from a real estate broker with "lots of affordable apartments," I was swindled out of hundreds of dollars in bogus fees. I was broke but I was determined to make it work.

New York definitely took some getting used to. The subway confused the hell out of me, so I would avoid transferring even if it meant walking thirty blocks to get to my destination. Hell, I figured, at least I was getting my exercise. Boy was I! I went from a size eight to a size seven, then from a seven to a size four, in just the first month. My doctor thought that I was stressing when I told her how much weight I had lost, but then when I told her how much I'd been walking, she realized that was the cause. I had her laughing when I told her, "The only thing I'm stressing is how I'm going to afford a new wardrobe!"

Finally, I found a place in Bedford-Stuyvesant, a.k.a., Bed-Stuy, a mostly low-income Caribbean neighborhood in Brooklyn that you always hear Lil' Kim rapping about. Everyone in that neighborhood had me pegged for a country girl the second I opened my mouth. "Say that again!" the guy behind the counter at KFC would shout, laughing, after I ordered my chicken dinner "wit a bisquit."

"You ain't from around here for sure!"

My living situation wasn't what I'd pictured, and neither was my job. People at POZ were very nice and welcoming, but the kind of work they had me doing was more than boring: emailing, returning phone calls, making spreadsheets. It was almost like they'd mixed up my titles; instead of ambassador, it seemed like I was everybody's administrative assistant.

"Unless I'm working for Diddy, holding that umbrella, I ain't no one's assistant," I thought, huffing through my tasks. (I had a big head from all my media exposure.) When I'd first met with the editor-in-chief she'd talked about me doing "development" work. What was I "developing"—returning other people's phone calls? But I had rent to pay and a new city to face. I couldn't just call someone in Nashville to come save me. I needed to deal. The savvy Marvelyn stepped up and gave the big head Marvelyn a talking to. I had to set my pride aside and get down to business. If returning phone calls was what I had to do, then I better become the best at returning phone calls and do it with a damn smile on my face.

The next few months I just went with the flow, got used to New York City, and ate more of those famous New York City hot dogs than I would like to remember because that was all my broke ass could afford. My rent was so high that after I paid that each month, I had no money left (not even for my shoe collection!). I made a few friends and started transferring to the local train so I didn't have to walk as far.

My patience at work was paying off. There were still some administrative tasks here and there, but I was starting to do more public speaking and media outreach—the kind of work where I could more immediately see the impact. I was featured in *Ebony* and on the cover of *The Ave*, an urban magazine. I spoke at a Washington Mystics game. I collaborated with various organizations such as the Black AIDS Institute and Youth AIDS. I spoke at the International AIDS Conference in Toronto and made a big impact there.

And then the grandmother of all opportunities came up . . . I got a phone call from *The Oprah Winfrey Show*. The producers had called every Brown in the Nashville phone book looking for me after seeing the article on me in *Newsweek*. They were doing an episode on HIV and AIDS and looking for women, in particular, with great personal stories and the personality to carry them off on national television.

The show went on to feature five other HIV-positive heterosexual women and me. Coming from around the country and from different ethnic backgrounds, we all had very different stories and experiences. The segment was a success. The show was titled "AIDS in America." I was sure it would make a difference in so many people's lives.

It was also my absolute honor—around this time—to be asked to do multiple speaking engagements for historically black colleges and universities in the south on World AIDS Day, December 1st. I knew that it was a significant opportunity to help the fastest growing population of new HIV cases in the United States. Today there are approximately 1.2 million people living with HIV in the United States, including 500,000 African Americans. Here was a chance for me to go home and speak my truth at colleges that I would have never been accepted to with my high school performance. It felt like I was returning to my roots, returning to people that I had, in a sense, run away from.

I started at Miles College in Alabama. The student response to my story was overwhelming. The health clinic on campus set up an HIV testing site afterward and they ran out of kits with about fifty people still in line. The next stop was Hampton University in Virginia, which was another success. The students called for HIV testing on their campus and I hear they got their wish.

Going to these historically black colleges and universities was so important to me. I was picked up not by a car service, but by people connected to the university who opened up their homes to me and welcomed me into their lives. It was really touching. I felt that the message and the personal connections were way more important than speaker's fees or car services, even if I was staying at a stranger's house.

Now and then my first HIV-positive love and I would chat on the phone, reminiscing about our wild and sex-filled tour, and swapping stories about speaking out along the way. I enjoyed our friendship, and only let myself fantasize that he would leave his girlfriend every once in awhile.

Then one day, I phoned and nonchalantly asked, "Hey, did you see me on *Oprah*?" hoping he would be impressed in the same way I used to be impressed by guys with flashy cars when I was nineteen.

"Yeah girl I seen you, you looked good," he said "If only I wasn't a married man . . ."

"Excuse me?" I said, having one of those this-can't-really-be-happening moments.

Sure enough, he repeated himself, "I got married a few weeks ago." I hung up and immediately deleted his number. I went through my entire phone book deleting numbers of all the men I'd hung on to. Prince Charming was the first to be axed.

Rejection was something I could never handle. So many of these men weren't rejecting my personality or my looks, they were rejecting HIV, even the one who had it. I thought that it would be easier to date someone who was positive, but even then there was stigma and the same heartache. I always wanted to be accepted. It's a great feeling when other people, especially the members of the opposite sex, tell you that you are beautiful and sexy, or that they love you. But I swear it is an even better feeling when you tell yourself that. I finally, truly, believed in that love. And I was starting to realize the only man in my life was God and truly He was all I needed.

That afternoon I headed to the tattoo parlor near my house and asked them to remake the guy's name on the small of my back into a butterfly. It hurt like hell, but it was an unforgettable symbol of the new Marvelyn. I wasn't going to wait for my knight in shining armor to convince me that I was lovable. It was my time to fly alone.

My twenty-third B-day party with my girls Sha,
Cortney, Kendria, and Furonda.
Photograph courtesy of Fred Cowan

chapter sixteen

While I was flying solo, I also felt the presence of angels.

I went home for the holidays and brought in the New Year with my nieces and my nephew—my real angels—in Nashville. We celebrated with sparkling grape juice, toasting to another year of health and growth.

They had all learned that I was HIV-positive back when the article in *The Tennessean* came out. Ever since then, each and every one of them asked me questions freely and I answered them openly. In fact, they had all become little peer educators of their own. Ask them a question, I double dare you; they know all the answers.

They've earned their awareness. My nieces and nephew endured name-calling and unflattering comments about their aunt at school when I took my story public. I never realized how much it would affect them. Here I was, trying to help people, but the side effect was that my loved ones suffered.

My conscience was eased when I realized that strength runs in the family. While at home, I noticed that my eleven-year-old niece had a red ribbon sticker on her bed. I said, "Jamiya, what does that mean?"

Without skipping a beat she said, "I support AIDS awareness."

And one time when I asked another niece, Diamond, "What have you learned from me?"

She replied, with a mischievous smile, "I'll be a virgin until the age of eighty, auntie."

But the youngest, the only boy, now he is something else. One day, not long after my diagnosis, he came into the room when I was in tears and asked me what was wrong. I told him that this guy did not want to talk to me because I had HIV. He looked me in the eyes and said, "If he don't want you because of HIV, he's missing out on a good person." He was nine years old at the time. I managed to look up and smile at him through the tears. These kids never saw me as just HIV. I was always auntie Marvy Marv to them. No matter what.

I got my inspiration from other HIV activists as well. Phill Wilson, an internationally known HIV/AIDS activist, invited me to speak at the It's All About Me Woman's Conference in Los Angeles—a real honor. Phill is currently the founder and Executive Director of the Black AIDS Institute, but has started so many HIV/AIDS organizations and education and research campaigns. He is also HIV-positive and has been living with the virus for twenty-five plus years and looks damn good. Most people are surprised when I tell them that I have been living with HIV for only five years, as if I was supposed to drop dead after the first 48 hours. Truth be told, AIDS has been around for more than three decades and people have been living with the virus for that long as well.

To me, Phill is a real role model because he doesn't just accept the status quo in AIDS education; instead he is constantly trying to innovate new approaches for a new generation of black youth. I hope to be constantly reinventing my work and myself in the same way that he has over the years, while still staying firmly dedicated to helping my people. Phill Wilson is a true survivor.

With Black AIDS Day approaching, February 7, along with the All About Me Conference, the speaking engagements all across the country were beginning to pour in. I went and spoke at Jackson State University in Mississippi and Tuskegee University in Alabama, two areas of the south that I was used to going to because of the lack of education and the infection rates.

During both programs, students swarmed the auditorium and asked questions, hungry for more information. It just seemed like no matter how much I got the word out, there were more people that needed to hear it. America thinks people know about HIV, but they don't. If they did, people wouldn't keep putting themselves at risk at such alarming rates every day.

After all that time back home in the south, I returned to Washington, D.C., to receive my Choice USA "Courage under Fire" award. All the speaking was beginning to die down and I had no extra money rolling in, which meant living paycheck to paycheck again, but I was excited to meet Gloria Steinem—famous feminist and the founder of Choice USA—and honored that the award had "courage" in the title. I couldn't claim to be a lot of things but I was one up on the lion from the Wizard of Oz. I had courage and I prided myself on that.

Despite my apparent angels, my commitment to keep going sometimes suffered. Despite the honors and the speaking opportunities, the hate mail kept flooding in:

"Just kill yourself already."

"If guys had any sense, they would see you and keep it moving!"

"You are glorifying AIDS!"

"You are walking bacteria!"

With all the ignorance flooding my email inbox and MySpace page, I began questioning my mission and my purpose. The fact that people were talking about me didn't affect me because that was oh-so-common. But I had to ask myself, are people not hearing my real message? Am I wasting my time? Why don't they feel that HIV can happen to them? Do they not realize that I do this for them? I know my status. I know about HIV. It affected me even more to know that I was just like them before my diagnosis.

I felt overwhelmed. Like most people, I had the stress of work, rent, bills, friendships, and a love life (or lack thereof). But unlike so many, I also had the unique stress that comes with being HIV-positive. Every day of my life I have to wake up and choose, once again, to live and live well. That means committing myself to taking my meds consistently, eating right, getting enough rest, and also being a role model for others struggling with my same predicament. I had put myself out there, thinking I was ready for the stigma and the discrimination. I'd experienced it so much even without going public, I figured it wouldn't be much different in the spotlight. But I was getting tired of all the responsibility. I just wanted to be twenty-two. Whatever in the hell that meant?

I mean, I was proud of what I had been doing, but I was burned out. I was doing things for other people—yes, I will switch planes in three different cities in one day to get an organization interested in HIV education the cheaper flight. Yes, I can come to your city on whatever day works best for you. Yes, yes, yes, really not thinking about my own health, and well-being.

And because I was in that place—doing for others and not myself—I felt as if people were ungrateful. I would ask myself, why in the hell do I care so much? Why am I trying to make a difference in the world that does not want my help? I realize now that I was reaching as many people as ever, but because I wasn't in the right state of mind, I was struggling. The means didn't seem to justify the ends. At the time, I felt ineffective. I decided I, at a minimum, needed a visit home.

Tab, my family, and my friends in Nashville threw me the biggest Sweet Twenty-third birthday party ever. We rented out a new restaurant on Jefferson Street and the theme was Hollywood. There was even a picture booth and a red carpet. When you entered the party there was a big three-foot poster of me from the cover that I did for *POZ* magazine.

I invited my old friend Furonda, from *America's Next Top Model*, and my closest friends and family. No one missed the chance to celebrate. With Kendria, Sha, and Cortney all there, it was like old times. To top it off, I had eight birthday cakes: one for each letter in my name.

All my people came out and showed me that they believed in what I was doing and that people actually did care. There were even people at my party who didn't initially want me to take my story public, who wanted me to live in silence, and were not educated themselves. I saw how these people, from my own community of Nashville, were changed and that meant the world to me. People needed this education and I had given it to them. I couldn't stop.

With Furonda in town and a power boost, my mom set up a speaking engagement with the all-girls track team that both LaRena and I had grown up on, the Continental T-Belles. Furonda and I spoke to the girls about the importance of loving yourself and never giving up on your dreams. It was meditation for me speaking with those girls.

My next adventure was going to Montego Bay, Jamaica. Even though I sure as hell needed one, it was not a vacation. There was

no sticking my feet in the sand. I was there to help end the HIV/
AIDS stigma in Jamaica—which was deeply rooted—by collabo-
rating with media moguls like I had in the United States.

I sat down with several HIV-positive Jamaicans and lis-
tened to their stories. The problems they were facing were acute;
HIV-positive people had been murdered, abused, and certainly
kicked out of families and fired from jobs. An HIV-positive man
that was in his early thirties told me that when he took his story
public only a few years earlier, he was shot, stabbed, and beaten
on multiple occasions. The pain these people endured put my
MySpace hate mail into perspective and made me outraged and
impassioned.

When I sat down with the writers and producers of the Carib-
bean Broadcast Network, they were shocked to learn that I was
HIV-positive. They actually had the nerve to ask if I was lying, if
I'd been paid to say that. Their faces looked totally confused.

I had to set them straight. The Caribbean is the second-most
HIV-affected region in the world, right after sub-Saharan Africa.
The message that needed, desperately, to reach both these execu-
tives and the millions of people they created television program-
ming for was not getting out. The media can be such an effective
way to combat ignorance and God knows there was a lot of it in
that beautiful paradise.

I didn't attack them like so many people had done me, but tried
to help them relate through my story. I gave them the hard facts
on HIV infection in their region—AIDS is now the leading cause
of death in many Caribbean countries; despite homophobia, the
most common method of transmission is through heterosexual
contact; with low condom use and 60 percent of the country under
the age of twenty-four, transmission rates are only expected to rise

if the Caribbean people don't take drastic measures to educate and protect themselves.

I also explained the nuts and bolts about just what HIV is and how it is passed from one person to the other. I talked about new innovations in medication that were allowing HIV-positive people to see their diagnosis, not as a death sentence, but as a serious wake-up call.

Then I charged them with a mission. "You have the power," I said, looking deep into their eyes. "As media executives, it is your duty to educate the people of the Caribbean about HIV and AIDS. You have the power to save lives, maybe even those of your family members or friends. No one is immune from this disease." After I was done telling my story and doing a little bit of preaching, one of the producers came up to me and said, "The white man brought that here." I have heard so many black people across the world say the same thing.

"Really?" I answered. I told him what I've told others: "Even if that were true, AIDS is something you have to acquire. You're telling me that you have no responsibility in that?" He just looked at me, stone cold, then said, "Thank you."

As I left the building and stepped into the hot air I thought, "This was way better than a vacation."

Flying back to New York City, I reflected on the past year. The party and my experience in Jamaica had put things firmly in perspective. I was making a difference. I was reaching people. I just needed to have a little more freedom to do that full time. The speaking was what made me the most fulfilled. It felt the most effective. When I returned to New York it was time to leave *POZ* with my head held high. It was time to be independent.

July 1, 2007, one year to the day after starting at *POZ*, I left. I didn't know where to begin, but I had myself, my family, my supporters, and God, and that is all I truly needed.

I started a company called Marvelous Connections. No, it is not a dating service; I cannot hook you up with a date. It is a consulting company, whose mission, is to offer an in-depth look into the world of HIV—no sugar coating—and provide people with a better understanding of a disease that is looked upon with fear and ignorance.

My first adventure as the founder and CEO of Marvelous Connections was to present at the NAACP Annual Convention this year held in Detroit, Michigan—a real honor considering the organization's rich history fighting discrimination and uplifting my people.

Next I had a chance to do more international work. Someone at the United States Embassy in Johannesburg, South Africa, saw an *Ebony* magazine article in which I was featured and decided I would be the right person to travel all around South Africa spreading a message of hope about living with HIV and AIDS and increasing the country's awareness of the real facts of transmission and infection (this is the country where President Thabo Mbeki actually questioned the link between HIV and AIDS, and claimed, despite leading a country where 5.4 million citizens are infected, that he did not know one HIV-positive person). I knew I had my work cut out for me.

I looked forward to going, because I knew that South Africa was more than the poverty I'd seen on TV. I was a bit concerned about what I would say, since South Africa has an infection rate five times that of the United States. I figured with those kinds of numbers I would be surrounded by experienced people who knew what I was dealing with on a daily basis, and hopefully, without so much stigma attached to it. Besides, as a child, I had always heard about AIDS in Africa. I figured that they got the memo. I was in for a big surprise.

From the second I stepped off that plane, South Africa provided the unexpected. First of all, with an estimated 350,000 South Africans dying from HIV-related causes each year, there is still a powerful HIV/AIDS stigma. HIV-positive people in this country were literally killing themselves shortly after leaving the testing clinic. The answer to the AIDS problem was peer silence. Three myths that South Africans believe to be true are, first, that AIDS can be cured by eating garlic potatoes; second, it can be cured by seeing a witch doctor; and, third, that if a person infected with HIV sleeps with a virgin, they can be cured. Young teenage men and women are being raped in the name of a so-called cure. I was shocked.

I was able to bring along my friend Sha, which gave us a rare opportunity to talk about how my HIV status had affected our friendship: not at all. You could tell that Sha's presence there, and even more, her clear acceptance of me, made the biggest impact on the crowds. Seeing me was wonderful. An audience member might think, "Okay, I can live with this." But seeing Sha meant, "I could even be loved with this." I imagined that made all the difference in the world to those people and I am eternally grateful to Sha for giving them that gift.

After my speeches, we were approached by two different kinds of women. Both told us that they really identified with the problem of craving the love of a man. Apparently that epidemic is global too! One group of women lived in the rural areas where they would be equivalent to the middle class in the United States. They talked about having sugar daddies who provided them with all the flash and bling. The other group of women was the total opposite. They lived in townships where there was no running water, no electricity, and children had to walk ten to twenty miles to go to school. They barely had clothing to cover their bony bodies and enough food to cover the table, while living six people to one tiny one-room house. They talked about having sugar daddies, not only for rent, but for food too. Bling was the last thing on their minds. As different as their lifestyles were, they had this in common—the sugar daddies talked these women into having unprotected sex and impregnated them, transmitted HIV to them, or both.

One woman who approached me truly struck a chord. Her name was Gugu. She told me that although she was HIV-positive, she had a healthy newborn baby girl. She continued to tell me that she struggled to make ends meet and was faced with a difficult decision. After seeing nine-year-old prostitutes sell their bodies for pennies throughout the country, I thought that would be her idea for some fast cash. But I was in for a rude awakening to learn she was struggling to choose whether to let her unborn baby starve to death or breastfeed her with the possibility of giving her HIV. It was a lot to swallow but I told her to stay strong and that her mothering instincts would kick in. In the meantime, I gave the woman some crumbled up crackers that were in the bottom of my purse. I had told Sha the day before I was not going to eat the crackers because they were not in the perfect square shape. Gugu smiled from ear to ear and said, "the happiness that you have just given me, I must give to my child."

Even in the face of all the serious, heartbreaking discussions, Sha and I managed to have fun while we were in South Africa. We did countless media appearances and posed for pictures with fans afterward. I swear I got at least a dozen serious marriage proposals. We stayed in beautiful hotels and ate at the fanciest restaurants. We were broke in the United States but in South Africa we were rich. We even snuck out one night—the trip was sponsored by PEPFAR, the President's Emergency Plan for AIDS Relief, established by George Bush in 2003, so we had highly protected armed vehicles and the whole nine—and went dancing at a club. Change the accents and a bit of the music and you would have never known that we were halfway around the world. Oh, except some guy at the club did tell me, "You move like Beyoncé." I never would have heard that in the United States. Maybe Keyshia Cole, but not Lady B.

We toured that whole country. We did so many interviews, speeches, and heard so many confessions. We also experienced hope. People were ready to make a change. People across the globe are still getting infected with a preventable disease. As soon as we blacks across the world start admitting that, the sooner that we as a people can get ourselves out from under this epidemic. South Africa was definitely a life changing experience for me, and I am very grateful that I took that twenty-three-hour plane trip.

As a black American woman with so many choices and opportunities, I found it devastating to see what women in developing countries must contend with. Medications aren't as easily available and the medical community isn't as informed as America's is. These women have fewer choices and less freedom than I do. Some of them cannot choose to go to school. Some of them cannot choose who they want to marry. Some of them cannot demand for a man to wear a condom. Some women don't even have the choice of becoming infected with HIV because they have been raped. My situation is a lot different from that of other black women around the world. Our choices are not the same. But one choice that we all have is the choice to love ourselves. We have to accept ourselves in whatever situation we are in, and know that we are living, beautiful people.

I know that now about myself. I have gone from an ignorant, needy girl in Nashville—desperate for love and looking in all the wrong places—to a strong, independent woman, taking care of myself financially, physically, and spiritually, and making the world a better place globally.

I have been through such vulnerability in my life—exposed for having a stigmatized disease, targeted, threatened, abandoned, even ridiculed. But through it all I had learned that the power of my life has nothing to do with others; it had everything to do with me, how I feel about myself, what I see in my future, the way I support and care for Marvelyn S. Brown.

I had nothing to hide. So when my friend and fellow HIV activist Jonathan Perry suggested I pose naked for a magazine spread that he and I were doing for *HIV Plus*, I eventually agreed. My gut reaction was "Hell no!" I had heard of too many girls who experienced a little bit of fame as permission to get naked. I valued my reputation and my privacy.

But with some reflection and conversation, I saw the more symbolic, deeper meaning in his artistic vision. He wanted me to show, with my bare skin, that my life was important—not because of the shoe collection or the Emmy or the celebrity meetings, not because of a man or his praise—but because it was mine, honest, strong, and plain. That I may be vulnerable at times, but at least I have no shame, no self-hate, no self-doubt left in me. I was the CEO not only of Marvelous Connections but also of Marvelyn's Life. The title of the article was "The Naked Truth," and I bared all.

I stood in front of the camera, naked as the day I was born, confident and proud. As the light fell on my dark skin, illuminating my form, I felt, finally, unequivocally, marvelous.

acknowledgments

Today I am very grateful for the relationships that I have formed with my family, myself, and God. The journey was not easy and I am happy that through the deepest and darkest time of my life, I had people that I could count on.

I am deeply grateful for the Kaiser Family, who gave me a platform to raise awareness about HIV to masses of young people. I am also grateful for some of their extremely hardworking partners that I had the pleasure of working with, including BET Rap-It-Up, MTV Think, and MTV International Staying-Alive. Thank you Tina Hoff, Jen Kates, Penny Duckham, Stephen Massey, Sonya Lockett, Vikki Johnson, Amy Campbell, Amy Brill, Ian Rowe, Georgia Arnold, Derek Wan, and Cathy Phiri.

I have had the opportunity to work with some very powerful organizations over the years. Special thanks to The Black AIDS Institute, The Diva Foundation, The Balm of Gilead, Youth AIDS, National Association of People With AIDS, NAACP Youth & College Division, The Magic Johnson Foundation, AIDS Alliance,

UNAIDS, New York AIDS Film Festival, Cable Positive, Choice USA, AIDS Alabama, and the National Coalition of 100 Black Women.

I would like to especially thank Nashville CARES for their dedication to middle Tennessee, and all AIDS service organizations and community-based organizations around the world. You all are my heroes.

To all of the media outlets who are not afraid to put HIV/AIDS information in their publications or programming, I truly thank you.

Thank you to my mother, my grandmother, my grandfather, my aunt Beverly, my uncles Del and Roger. My marvelous sisters Tab and Mone't. Thank you Diamond, Jamiya, and Jamarius for always showing me the hottest dance steps and the latest slang. I am especially thankful to my dedicated cousins: Stan, Martina, Jasmine, Stacy, Mario, Alex, and Jermar.

Thank you to all my friends: Sha, Kendria, Ashley, Netta, Henry, Chelsea, Johnny, Jeff, Jonathan, Laura, Diana, CaTina, Ambur, Alesha Renee, Charlotte, Peter, Cassie, Delon, Tim, Lynnea, Mike, Kellee, Shatoya, Shunta, Laurel, Furonda, Herk, Heather, Patriz, Nicole, Carol, Sammie, Patrick, Zion, and Deaundre. Thank you for wiping my tear and celebrating my every cheer. That was cheesy, but I mean it.

Thank you to my many mentors throughout the years for your honesty and realness: Coach Smith, Sydney Hardaway, Jessica Poyner, Tresa Reigart, Mean Strubb, Lynn Chamberlin, Patrick Luther, Gloria Rueben, Stephanie Fredric, Suzanne Africa Engo, and Andrew Spieldenner. Your great spirits stay with me.

Thanks to my case manager Marcia Williams, my doctors Lisa Laya and Beverly Byram, and my perky little nurse, Sam. Special thanks to Yvette Moore. You guys rock. Thank you to the lovely people who work at Gilead Pharmaceuticals and the Community Education Task Force members. Special thanks to James Loduca, Deborah Wafer, and Elaine Ray.

Thank you to the soulful voices and lyrics of Alicia Keys, Common, John Legend, Kanye West, and Fantasia for helping me through some of the most challenging parts of my life.

Thank you to Jamie Brickhouse and the HarperCollins Speaker's Bureau for handling all my speaking engagements. I am grateful for Tracy Brown, my literary agent, for connecting me with Amistad/HarperCollins, and thanks to my marvelous writer, Courtney E. Martin, for putting my story into words. Thank you to Dawn Davis, Bryan Christian, Gilda Squire, Adrienne Rhodes, and the rest of the team at Amistad/HarperCollins who have embraced this book and my story with such excitement. Thanks also to Gil Robertson IV for his insight on writing a book.

To all the people that let my words inspire them to take control of their lives and their futures, I do this for you. To those who are still ignorant, opinionated, and have negative things to say, you show me that my work is far from done.

If I have forgotten anyone, please remember that I take HIV medication and one of the side effects is short-term memory loss. Thank you, and I love you all.

appendix A: statistics

The statistics below make for some almost unbelievable reading, but they are all true. The scariest thing is that these numbers keep on increasing. It is up to us—each and every one of us—to stop this disease.

UNITED STATES

- An estimated 1 million people are living with HIV/AIDS in the United States.
- At least 40,000 people are infected each year.
- African Americans account for more than half of new HIV infections.
- AIDS is the leading cause of death for African American women aged 25 to 34.
- African Americans account for more HIV and AIDS cases and HIV-related deaths than any other racial group in the United States.
- Hispanics are the second most likely to suffer.

Source: Youth AIDS (www.youthAIDS.org)

GLOBAL

- More than 42 million people now live with HIV or AIDS.
- Every day 14,000 people contract HIV—that's 10 people per minute.
- Every 10 seconds someone dies of AIDS-related illness.
- One child dies of AIDS-related illness every minute.
- Losing one or both parents to AIDS, more than 15 million children around the world are AIDS orphans.
- Every 15 seconds another person between the ages of 15 and 24 becomes infected with HIV.

Source: Youth AIDS (www.youthAIDS.org)

appendix B: five common myths about HIV/AIDS

THE MYTH: I'm safe because I'm a virgin.

THE REALITY: Virgin is only a label. If you have had oral or anal sex but never vaginal sex, you are at risk. Even if you have had no sexual contact at all, you are not immune. Sex is only *one* mode of transmission. HIV can be transmitted through sharing needles (tattoos, body piercings, or IV drug use) and from mother to child through breast-feeding or during the delivery process.

THE MYTH: HIV can be spread through tears, sweat, mosquitoes, pools, or casual contact.

THE REALITY: HIV can only be transmitted through infected blood, semen, vaginal fluids, and breast milk. HIV can also be passed from mother to baby.

THE MYTH: Straight people don't get HIV.

THE REALITY: The majority of HIV-positive people worldwide are heterosexual. Risk is not about labeling people; it is about behavior.

THE MYTH: I'm safe because I'm in a monogamous relationship or married.

THE REALITY: Unfortunately, a wedding ring or a monogamous relationship cannot protect you from HIV risk. If you're faithful but your partner is not, or your partner was already HIV-positive before you met, you can still get HIV. You should ask yourself these questions: Were you tested for HIV before you got into the relationship? Was your partner? Were both tests negative? Do you spend twenty-four hours a day together?

THE MYTH: Lesbians don't get HIV.

THE REALITY: Women who have sex only with women are generally at lower risk because of the sexual activities they engage in, but they can still get HIV.

HIV stands for "human immunodeficiency virus." *AIDS* stands for "acquired immunodeficiency syndrome." HIV is the virus that causes the disease AIDS.

HIV is spread through the following bodily fluids: blood, semen, vaginal fluids, and breast milk.

HIV is passed from one person to another by
- Having sex (vaginal, anal, or oral) with a person who has HIV
- Sharing needles with a drug user who has HIV
- During pregnancy, birth, or breast-feeding if a mother has HIV
- Through transfusions of blood with HIV, which is now rare in the United States but still happens in other countries

appendix D: websites and hotlines

WEBSITES

Health-related
www.thebody.com

Relationships and sexual health
www.drrachael.com

Policy-related
www.kff.org

Testing (find a site near you)
www.hivtest.org

Take action
www.think.mtv.com

HIV and youth

www.youthAIDS.org

HIV and African Americans

www.rapituppresents.com

HIV and Latinos

www.latinoAIDS.org

AIDS COMMUNITIES ONLINE

Poz I Am

www.poziam.com

My HIV Life

www.myhivlife.com

TOLL-FREE HOTLINES

BET RAP IT UP

1-866-RAP-IT-UP

Information about HIV, STDs, and pregnancy

Centers for Disease Control

1-800-342-2437

National AIDS hotline

Centers for Disease Control in Spanish

1-800-344-7432

National AIDS hotline in Spanish/Español

discussion guide
*(for use in classrooms, book groups,
or between friends and family members)*

1. The story of Marvelyn's parents' divorcing is an all-too-common reality in this country. Do you see that as a trend that will change any time soon, or is divorce just a reality of contemporary life?

2. Clearly Marvelyn and her sister, Mone't, had different talents. So often siblings can feel pitted against one another or typecast in a certain way. Did you experience this as a child? How do you think it influenced your self-image as you grew up?

3. Marvelyn's mom seems to be the kind of woman who subscribes to tough love. What do you think about this approach to parenting? Did you have a tough-love parent? If so, how did he or she make you feel about yourself?

4. Marvelyn appears to be typecast, once again, among her group of high school friends. Did you experience this as a teenager?

In what ways did it support you and in what ways was it limiting?

5. What kind of sex education did you receive? In school? From family or friends? What do you wish you had been told? What kind of messages about sexually transmitted diseases did you receive?

6. Marvelyn considers suicide. The suicide rate has been rising—for the first time in decades—among American teenagers. Why do you think this is? What can we do about it?

7. Marvelyn has trouble finding reliable birth control methods that she's comfortable with. What kinds of messages do you see in the media, at home, or in school about contraception? What's missing?

8. Marvelyn experiences a lot of violence throughout her high school years. Why do you think teenagers resort to physical violence? What could she have done differently in the situations she faced?

9. During her later teen years, we see a trend in Marvelyn's love life. She seems very impressionable and hungry for a boyfriend. Why do you think this is? Why do so many girls feel like they *need* a boyfriend? What larger need might it be masking?

10. Marvelyn seems to have little idea of where her life is headed after graduation. Why do you think this is? If you're past this

life stage, look back. Did you have a strong sense of what you wanted to do with your life? If so, where did that come from? If not, what would have given you that direction?

11. How do you see Marvelyn's self-image changing after high school? Do you think it is a change for the good? Why or why not?

12. LaRena, Marvelyn's friend, dies so tragically. Have you ever experienced the suicide of a close friend or loved one? What was that like? How did it change your ideas about your own life?

13. Marvelyn is deeply disappointed in her ACT score. When have you experienced that kind of disappointment, and how did you deal with it?

14. Marvelyn thinks she's met Prince Charming. What really characterizes a healthy relationship? What must be present after the "honeymoon period" is over? What did you think about her interpretation of her boyfriend's willingness to have sex without a condom—that it meant he really loved her—and her ultimate choice to have unprotected intercourse?

15. What have you been taught about HIV and AIDS? Do you see yourself as vulnerable to the disease? If you already have it, what was it like to read Marvelyn's story of being diagnosed?

16. Were you surprised by how little Marvelyn knew about HIV and AIDS? Why are so many in this country—and beyond—still so ignorant about the disease?

17. To your mind, who is responsible for Marvelyn's infection? Prince Charming? Marvelyn? The educators who didn't fully inform her about the disease? Her family? Her church?

18. Why do you think Marvelyn went back to Prince Charming after she learned he had infected her with HIV?

19. What are the deeper reasons, in your opinion, that Marvelyn's local community responded so badly to her diagnosis? How has or would your community respond? How would you respond if you heard your best friend or sibling had HIV?

20. Why did Marvelyn refuse to disclose Prince Charming's identity? Do you respect her choice? Why or why not?

21. How did Marvelyn's near car crash change her outlook on life? Have you ever had a similar transformative moment?

22. What do you think about Marvelyn's decision to go public about her disease? What would you have done in her situation?

23. What are the most glamorous parts of Marvelyn's ascent to being a public figure? What are the hardest parts?

24. What role do men play in Marvelyn's continuing struggle to find self-love and make responsible choices? Consider her father and her boyfriends, both HIV negative and positive.

25. Do you agree with Marvelyn that too many women are lacking in self-love? Why is this lack so epidemic among contemporary women?

26. Marvelyn realized that "with power comes responsibility." When have you experienced this in your life? How did you deal with it?

27. How do outlooks on HIV differ between Jamaica, South Africa, and the United States?

28. How will reading this book change your outlook and behavior? What was most surprising? Most inspiring?

29. Who is one other person you really want to read this book and why?

Contact Marvelyn with any of your own questions at www.marvelynbrown.com. She'd love to hear from you.